U.S. Department of Transportation
National Highway Traffic Safety Administration

DOT HS 810 757

April 2007

Development of Crash Imminent Test Scenarios for Integrated Vehicle-Based Safety Systems (IVBSS)

This document is available to the public from the National Technical Information Service, Springfield, Virginia 22161

This publication is distributed by the U.S. Department of Transportation, National Highway Traffic Safety Administration, in the interest of information exchange. The opinions, findings and conclusions expressed in this publication are those of the author(s) and not necessarily those of the Department of Transportation or the National Highway Traffic Safety Administration. The United States Government assumes no liability for its content or use thereof. If trade or manufacturer's names or products are mentioned, it is because they are considered essential to the object of the publication and should not be construed as an endorsement. The United States Government does not endorse products or manufacturers.

REPORT DOCUMENTATION PAGE

Form Approved
OMB No. 0704-0188

Public reporting burden for this collection of information is estimated to average one hour per response, including the time for reviewing instructions, searching existing data sources, gathering and maintaining he data needed, and completing and reviewing the collection of information. Send comments regarding this burden estimate or any other aspect of this collection of information, including suggestions for reducing this burden, to Washington Headquarters Services, Directorate for Information Operations and Reports, 1215 Jefferson Davis Highway, Suite 1204, Arlington, VA 22202-4302, and to the Office of Management and Budget, Paperwork Reduction Project (0704-0188), Washington, DC 20503.

1 AGENCY USE ONLY (Leave blank)	2. REPORT DATE April 2007	3. REPORT TYPE AND DATES COVERED Final Report October 2005 – December 2006
4. TITLE AND SUBTITLE Development of Crash Imminent Test Scenarios for Integrated Vehicle-Based Safety Systems		5. FUNDING NUMBERS PPA# HS22 WPB# DG025
6. AUTHOR(S) Wassim G. Najm and John D. Smith		
7. PERFORMING ORGANIZATION NAME(S) AND ADDRESS(ES) U.S. Department of Transportation Research and Innovative Technology Administration Advanced Safety Technology Division John A. Volpe National Transportation Systems Center Cambridge, MA 02142		8. PERFORMING ORGANIZATION REPORT NUMBER DOT-VNTSC-NHTSA-07-01
9. SPONSORING/MONITORING AGENCY NAME(S) AND ADDRESS(ES) U.S. Department of Transportation National Highway Traffic Safety Administration		10. SPONSORING/MONITORING AGENCY REPORT NUMBER DOT HS 810 757
11. SUPPLEMENTARY NOTES		
12a. DISTRIBUTION/AVAILABILITY STATEMENT This document is available to the public through the National Technical Information Service, Springfield, Virginia 22161.		12b. DISTRIBUTION CODE

13. ABSTRACT (Maximum 200 words)

This report identifies crash imminent test scenarios based on common pre-crash scenarios for integrated vehicle-based safety systems that alert the driver of a light vehicle or a heavy truck to an impending rear-end, lane change, or run-off-road crash. Pre-crash scenarios describe vehicle movements and critical events immediately prior to the crash. The General Estimates System (GES) crash database was queried to distinguish common pre-crash scenarios for light vehicles (2003 GES) and heavy trucks (2000-2003 GES) in terms of their frequency of occurrence. Analysis of two-vehicle rear-end crashes revealed four dominant scenarios that accounted for 97 percent of light-vehicle crashes and 95 percent of heavy-truck crashes in which the subject vehicle was striking. Four scenarios were also identified from an analysis of two-vehicle lane change crashes, comprising 65 percent of light-vehicle crashes and 76 percent of heavy-truck crashes in which the subject vehicle was encroaching onto another vehicle in adjacent lanes. There were five single-vehicle, run-off-road scenarios representing 63 percent of light-vehicle crashes and 83 percent of heavy-truck crashes, excluding crashes caused by vehicle failure or evasive maneuver. An additional set of scenarios is proposed to address multiple threats from near simultaneous critical events. This report also provides a statistical description of individual scenarios in terms of their environmental factors, roadway geometry, and speed conditions.

14. SUBJECT TERMS Integrated vehicle-based safety systems, crash imminent test scenarios, pre-crash scenarios, rear-end crash, lane change crash, run-off-road crash, multiple-threat scenarios, light vehicle, and heavy truck.			15. NUMBER OF PAGES 57
			16. PRICE CODE
17. SECURITY CLASSIFICATION OF REPORT Unclassified	18. SECURITY CLASSIFICATION OF THIS PAGE Unclassified	19. SECURITY CLASSIFICATION OF ABSTRACT Unclassified	20. LIMITATION OF ABSTRACT

NSN 7540-01-280-5500

Standard Form 298 (Rev. 2-89)
Prescribed by ANSI Std. 239-18
298-102

METRIC/ENGLISH CONVERSION FACTORS

ENGLISH TO METRIC | METRIC TO ENGLISH

LENGTH (APPROXIMATE)

English to Metric	Metric to English
1 inch (in) = 2.5 centimeters (cm)	1 millimeter (mm) = 0.04 inch (in)
1 foot (ft) = 30 centimeters (cm)	1 centimeter (cm) = 0.4 inch (in)
1 yard (yd) = 0.9 meter (m)	1 meter (m) = 3.3 feet (ft)
1 mile (mi) = 1.6 kilometers (km)	1 meter (m) = 1.1 yards (yd)
	1 kilometer (km) = 0.6 mile (mi)

AREA (APPROXIMATE)

English to Metric	Metric to English
1 square inch (sq in, in^2) = 6.5 square centimeters (cm^2)	1 square centimeter (cm^2) = 0.16 square inch (sq in, in^2)
1 square foot (sq ft, ft^2) = 0.09 square meter (m^2)	1 square meter (m^2) = 1.2 square yards (sq yd, yd^2)
1 square yard (sq yd, yd^2) = 0.8 square meter (m^2)	1 square kilometer (km^2) = 0.4 square mile (sq mi, mi^2)
1 square mile (sq mi, mi^2) = 2.6 square kilometers (km^2)	10,000 square meters (m^2) = 1 hectare (ha) = 2.5 acres
1 acre = 0.4 hectare (he) = 4,000 square meters (m^2)	

MASS - WEIGHT (APPROXIMATE)

English to Metric	Metric to English
1 ounce (oz) = 28 grams (gm)	1 gram (gm) = 0.036 ounce (oz)
1 pound (lb) = 0.45 kilogram (kg)	1 kilogram (kg) = 2.2 pounds (lb)
1 short ton = 2,000 pounds (lb) = 0.9 tonne (t)	1 tonne (t) = 1,000 kilograms (kg) = 1.1 short tons

VOLUME (APPROXIMATE)

English to Metric	Metric to English
1 teaspoon (tsp) = 5 milliliters (ml)	1 milliliter (ml) = 0.03 fluid ounce (fl oz)
1 tablespoon (tbsp) = 15 milliliters (ml)	1 liter (l) = 2.1 pints (pt)
1 fluid ounce (fl oz) = 30 milliliters (ml)	1 liter (l) = 1.06 quarts (qt)
1 cup (c) = 0.24 liter (l)	1 liter (l) = 0.26 gallon (gal)
1 pint (pt) = 0.47 liter (l)	
1 quart (qt) = 0.96 liter (l)	
1 gallon (gal) = 3.8 liters (l)	
1 cubic foot (cu ft, ft^3) = 0.03 cubic meter (m^3)	1 cubic meter (m^3) = 36 cubic feet (cu ft, ft^3)
1 cubic yard (cu yd, yd^3) = 0.76 cubic meter (m^3)	1 cubic meter (m^3) = 1.3 cubic yards (cu yd, yd^3)

TEMPERATURE (EXACT)

English to Metric	Metric to English
[(x-32)(5/9)] °F = y °C	[(9/5) y + 32] °C = x °F

QUICK INCH - CENTIMETER LENGTH CONVERSION

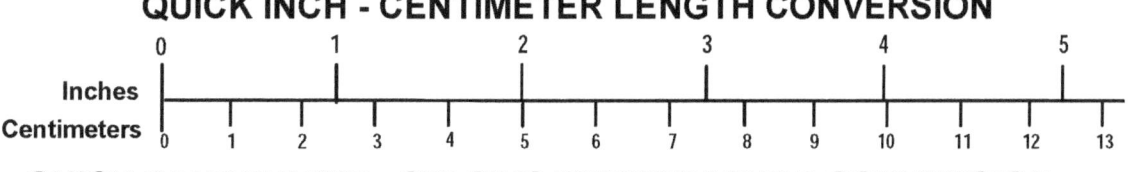

QUICK FAHRENHEIT - CELSIUS TEMPERATURE CONVERSION

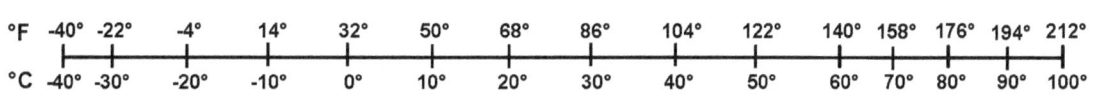

For more exact and or other conversion factors, see NIST Miscellaneous Publication 286, Units of Weights and Measures. Price $2.50 SD Catalog No. C13 10286

Updated 6/17/98

PREFACE

The Volpe National Transportation Systems Center (Volpe Center) of the United States Department of Transportation's Research and Innovative Technology Administration is conducting an independent evaluation of integrated safety systems for motor vehicles in support of the National Highway Traffic Safety Administration (NHTSA). This research activity represents a part of the Integrated Vehicle-Based Safety Systems (IVBSS) initiative in the Intelligent Transportation Systems (ITS) program. The goal of IVBSS is to accelerate the deployment of integrated crash warning systems for passenger cars and heavy commercial trucks to prevent rear-end, lane change, and run-off-road crashes.

This report presents the results obtained for the analysis of rear-end, lane change, and run-off-road crashes based on statistics from the 2000-2003 National Automotive Sampling System General Estimates System crash databases. In 2003, there were approximately 3,822,000 police-reported (PR) rear-end, lane change, and off-roadway crashes involving all vehicle types in the United States. These three crash types accounted for about 60 percent of all PR crashes.

The authors of this report are Wassim Najm and John Smith of the Volpe Center.

The authors acknowledge the technical contribution of Samuel Toma and Mikio Yanagisawa of the Volpe Center for processing the data and setting up the tables used in the report. Also acknowledged are Jack Ference of NHTSA, Sandor Szabo of the National Institute of Standards and Technology (NIST), and other reviewers from NHTSA, NIST, Noblis (formerly Mitretek), and SAIC for reviewing the report and providing valuable comments.

TABLE OF CONTENTS

Section   **Page**

- LIST OF ACRONYMS vi
- LIST OF ABBREVIATIONS FOR CRASH IMMINENT SCENARIOS vii
- EXECUTIVE SUMMARY viii
- 1. INTRODUCTION 1
- 2. LIGHT-VEHICLE SCENARIOS 3
  - 2.1. Rear-End Crash Imminent Test Scenarios 3
  - 2.2. Lane Change Crash Imminent Test Scenarios 5
  - 2.3. Run-Off-Road Crash Imminent Test Scenarios 7
- 3. HEAVY-TRUCK SCENARIOS 11
  - 3.1. Rear-End Crash Imminent Test Scenarios 11
  - 3.2. Lane Change Crash Imminent Test Scenarios 13
  - 3.3. Run-Off-Road Crash Imminent Test Scenarios 15
- 4. MULTIPLE-THREAT CRASH IMMINENT TEST SCENARIOS 19
  - 4.1. Rear-End and Lane Change Crash Imminent Test Scenarios 19
  - 4.2. Lane Change and Run-Off-Road Crash Imminent Test Scenarios 19
  - 4.3. Rear-End and Run-Off-Road Crash Imminent Test Scenarios 20
- 5. RECOMMENDED CRASH IMMINENT BASE TEST SCENARIOS 21
- 6. CONCLUSIONS 26
- 7. REFERENCES 27
- APPENDIX A. Identification of Dynamically Distinct Pre-Crash Scenarios Using GES Codes 28
  - Vehicle Type Identification 28
  - Rear-End Pre-Crash Scenarios 28
  - Lane Change Pre-Crash Scenarios 30
  - Run-Off-Road Pre-Crash Scenarios 31
- APPENDIX B. Statistical Description of Base Test Scenarios 33
  - Light Vehicle Statistics Based on 2003 GES 33
  - Heavy-Truck Statistics Based on 2000-2003 GES 35
- APPENDIX C. First Harmful Event Statistics of Run-Off-Road Crash Scenarios 37
  - Light-Vehicle Statistics Based on 2003 GES 37
  - Heavy Truck Statistics Based on 2000-2003 GES 41

LIST OF TABLES

Table — **Page**

Table 1. Frequency Distribution of Light-Vehicle Rear-End Pre-Crash Scenarios 4
Table 2. Recommended Light-Vehicle, Rear-End Crash Imminent Base Test Scenarios 5
Table 3. Frequency Distribution of Light-Vehicle Lane Change Pre-Crash Scenarios 6
Table 4. Breakdown of Light-Vehicle Turning Scenarios by Movement of Other Vehicle 6
Table 5. Recommended Light-Vehicle, Lane Change Crash Imminent Base Test Scenarios 7
Table 6. Frequency Distribution of Single Light-Vehicle Run-Off-Road Pre-Crash Scenarios ... 8
Table 7. Trafficway Flow Information for Light-Vehicle Left-Road-Edge Departure 8
Table 8. First Harmful Event Statistics of Single Light-Vehicle Run-Off-Road Crashes 10
Table 9. Recommended Light-Vehicle, Run-Off-Road Crash Imminent Base Test Scenarios .. 10
Table 10. Frequency Distribution of Heavy Truck Rear-End Pre-Crash Scenarios 11
Table 11. Recommended Heavy Truck, Rear-End Crash Imminent Base Test Scenarios 12
Table 12. Frequency Distribution of Heavy-Truck Lane Change Pre-Crash Scenarios 13
Table 13. Breakdown of Heavy-Truck Turning Scenarios by Movement of Other Vehicle 14
Table 14. Recommended Heavy-Truck, Lane Change Crash Imminent Base Test Scenarios 15
Table 15. Frequency Distribution of Heavy-Truck Run-Off-Road Pre-Crash Scenarios 15
Table 16. Trafficway Flow Information for Heavy-Truck Left-Road-Edge Departure 16
Table 17. First Harmful Event Statistics of Heavy-Truck Run-Off-Road Crashes 17
Table 18. Recommended Heavy Truck, Run-Off-Road Crash Imminent Base Test Scenarios . 18
Table 19. Recommended Rear-End Crash Imminent Base Test Scenarios for Light Vehicle and Heavy Truck 22
Table 20. Recommended Lane Change Crash Imminent Base Test Scenarios for Light Vehicles and Heavy Trucks 23
Table 21. Recommended Run-Off-Road Crash Imminent Base Test Scenarios for Light Vehicles and Heavy Trucks 24
Table 22. Recommended Multiple-Threat Base Test Scenarios for Light Vehicles and Heavy Trucks 25
Table 23. Driving Environment of Light Vehicle Rear-End Crash Base Test Scenarios 33
Table 24. Speed Data of Light-Vehicle Rear-End Crash Base Test Scenarios 33
Table 25. Driving Environment of Light-Vehicle Lane Change Crash Base Test Scenarios 33
Table 26. Speed Data of Light-Vehicle Lane Change Crash Base Test Scenarios 34
Table 27. Driving Environment of Light-Vehicle Run-Off-Road Crash Base Test Scenarios ... 34
Table 28. Speed Data of Light-Vehicle Run-Off-Road Crash Base Test Scenarios 34
Table 29. Driving Environment of Heavy-Truck Rear-End Crash Base Test Scenarios 35
Table 30. Speed Data of Heavy-Truck Rear-End Crash Base Test Scenarios 35
Table 31. Driving Environment of Heavy-Truck Lane Change Crash Base Test Scenarios 35
Table 32. Speed Data of Heavy-Truck Lane Change Crash Base Test Scenarios 36
Table 33. Driving Environment of Heavy-Truck Run-Off-Road Crash Base Test Scenarios 36
Table 34. Speed Data of Heavy-Truck Run-Off-Road Crash Base Test Scenarios 36
Table 35. Light-Vehicle First Harmful Events: Going Straight and Departing Road Edge Scenarios 37
Table 36. Light-Vehicle First Harmful Events: Negotiating a Curve and Departing Road Edge Scenarios 38

Table 37. Light-Vehicle First Harmful Events: Control Loss Scenarios 39
Table 38. Light-Vehicle First Harmful Events: Initiating a Maneuver Scenarios 40
Table 39. Heavy-Truck First Harmful Events: Going Straight and Departing Road Edge Scenarios 41
Table 40. Heavy-Truck First Harmful Events: Negotiating a Curve and Departing Road Edge Scenarios 41
Table 41. Heavy-Truck First Harmful Events: Control Loss Scenarios 42
Table 42. Heavy-Truck First Harmful Events: Initiating a Maneuver Scenarios 42

LIST OF ACRONYMS

GES	General Estimates System
ITS	Intelligent Transportation Systems
IVBSS	Integrated Vehicle-Based Safety Systems
PR	Police-Reported
PSU	Primary Sampling Unit
ROR	Run-Off-Road
USDOT	United States Department of Transportation

LIST OF ABBREVIATIONS FOR CRASH IMMINENT SCENARIOS

- **A1**   Light vehicle makes a maneuver and encounters a lead vehicle
- **A2**   Light vehicle encounters a lead vehicle moving at a lower constant speed
- **A3**   Light vehicle encounters a decelerating lead vehicle
- **A4**   Light vehicle encounters a stopped lead vehicle
- **B1**   Light vehicle changes lanes or passes to the right and encroaches on adjacent vehicle
- **B2**   Light vehicle changes lanes or passes to the left and encroaches on adjacent vehicle
- **B3**   Light vehicle turns left and encroaches on adjacent vehicle
- **B4**   Light vehicle drifts right and encroaches on adjacent vehicle
- **C1**   Light vehicle is going straight and departs road edge to the right
- **C2**   Light vehicle is going straight and departs road edge to the left
- **C3**   Light vehicle is negotiating a curve and departs road edge to the right
- **C4**   Light vehicle is negotiating a curve and loses control
- **C5**   Light vehicle is initiating a maneuver and departs road edge to the right
- **D1**   Heavy truck makes a maneuver and encounters a lead vehicle
- **D2**   Heavy truck encounters a lead vehicle moving at a lower constant speed
- **D3**   Heavy truck encounters a decelerating lead vehicle
- **D4**   Heavy truck encounters a stopped lead vehicle
- **E1**   Heavy truck changes lanes or passes to the right and encroaches on adjacent vehicle
- **E2**   Heavy truck changes lanes or passes to the left and encroaches on adjacent vehicle
- **E3**   Heavy truck turns right and encroaches on adjacent vehicle
- **E4**   Heavy truck drifts right and encroaches on adjacent vehicle
- **F1**   Heavy truck is going straight and departs road edge to the right
- **F2**   Heavy truck is going straight and departs road edge to the left
- **F3**   Heavy truck is negotiating a curve and departs road edge to the right
- **F4**   Heavy truck is negotiating a curve and loses control
- **F5**   Heavy truck is initiating a maneuver and departs road edge to the right

EXECUTIVE SUMMARY

This report recommends a basic set of crash imminent test scenarios for integrated vehicle-based safety systems designed to warn the driver of an impending rear-end, lane change, or run-off-road crash. Four crash threat types are considered:

- Rear-end crash
- Lane change crash
- Run-off-road crash
- Multiple threats (Combinations of rear-end, lane change, and run-off-road crash threats)

The Integrated Vehicle-Based Safety Systems (IVBSS) initiative of the USDOT Intelligent Transportation Systems (ITS) program will build and field test integrated system prototypes for light vehicles (e.g., passenger cars, vans, minivans, sport utility vehicles, and light pickup trucks) and heavy trucks (i.e., gross vehicle weight rating over 10,000 pounds). This initiative seeks to accelerate the deployment of integrated systems that warn drivers when the vehicle they are driving: 1) is about to strike a slower-moving vehicle traveling ahead of them in the same lane and direction of travel; 2) is about to strike a vehicle that is stopped in their travel lane; 3) is leaving the roadway or traveling too fast for an upcoming curve; or 4) is about to collide with a vehicle in an adjacent lane when the driver is executing a lane change maneuver.

The IVBSS initiative will develop objective test procedures for light vehicles and heavy trucks, including crash imminent test scenarios and operational scenarios (e.g., "do not warn" scenarios to minimize false alerts). Objective test procedures are important to verify that IVBSS prototypes meet their performance specifications and are safe for use by ordinary drivers prior to the start of field operational tests. Crash imminent test scenarios are based on crash data from crash types targeted by IVBSS. Operational scenarios are devised from the capability and limitations of the state-of-the-art technologies in resolving targets and environmental conditions encountered in real-world driving. Objective test procedures will provide a complete set of:

- Initial kinematic conditions for each test scenario;
- Driving conditions;
- Instructions on how to run each scenario;
- Test apparatus (instrumentations and props);
- Measures of performance; and
- Pass-fail criteria.

This report focuses on crash imminent test scenarios and recommends a set of scenarios based on most common pre-crash scenarios as identified from the General Estimates System (GES) crash database. These scenarios represent the majority of driving conflicts that IVBSS functions should address on public roadways. This report statistically describes individual scenarios in terms of their environmental factors (weather and lighting conditions), roadway geometry (alignment and profile), and speed conditions (posted speed limit and speeding information). This report also suggests a list of crash imminent test scenarios that examine the capability of the integrated crash warning system in dealing with multiple-threat scenarios.

Test scenarios are provided below for rear-end, lane change, and run-off-road crash countermeasures along with the annual frequency of occurrence of their respective pre-crash scenarios based on 2003 GES data for light vehicles and 2000-2003 GES data for heavy trucks. Multiple years of GES data were needed to better represent pre-crash scenarios of heavy trucks. The following scenarios were developed:

Rear-End Crash Imminent Test Scenarios

Light vehicles and heavy trucks were involved in about 1,483,000 and 64,000 police-reported (PR) two-vehicle rear-end crashes, respectively. The light vehicle and heavy truck rear-ended a lead vehicle in almost 95 percent and 60 percent of these crashes, respectively. The host vehicle below refers to the light vehicle or heavy truck to be equipped with IVBSS. The following four crash imminent scenarios are recommended as base test scenarios for the rear-end crash warning function:

1. Host vehicle changes lanes (light vehicle at 35-60 mph and heavy truck at 35-55 mph) and encounters a stopped lead vehicle ahead in daylight, clear weather, on straight and level road – light vehicle and heavy truck made a maneuver (e.g., lane change) and struck a lead vehicle in 75,000 and 2,000 rear-end crashes per year, respectively.
2. Host vehicle is moving at constant speed (light vehicle at 45-60 mph and heavy truck at 45-55 mph) and encounters a lead vehicle moving at slower constant speed in daylight, clear weather, on straight and level road – light vehicle and heavy truck struck a lead vehicle moving at lower constant speed in 200,000 and 8,000 rear-end crashes, respectively.
3. Host vehicle is following a lead vehicle at constant speed (light vehicle at 45-60 mph and heavy truck at 35-55 mph) and then lead vehicle suddenly decelerates in daylight, clear weather, on straight and level road – light vehicle and heavy truck struck a decelerating lead vehicle in 730,000 and 13,000 rear-end crashes, respectively.
4. Host vehicle is moving at constant speed (light vehicle and heavy truck at 35-55 mph) and encounters a stopped lead vehicle in daylight, clear weather, on straight and level road – light vehicle and heavy truck struck a stopped lead vehicle in 364,000 and 12,000 rear-end crashes, respectively.

Lane Change Crash Imminent Test Scenarios

Light vehicles and heavy trucks were involved in 544,000 and 78,000 PR two-vehicle lane change crashes, respectively. The light vehicle and heavy truck encroached into the adjacent lane of other vehicle types in about 82 percent and 59 percent of these crashes, respectively. The following four crash imminent scenarios are recommended as base test scenarios for the lane change crash warning function:

1. Host vehicle changes lanes to the right (constant longitudinal speed at 35-60 mph for light vehicles and 35-55 mph for heavy trucks) and encroaches on an adjacent vehicle in daylight, clear weather, on straight and level road – light vehicle and heavy truck collided

with an adjacent vehicle during a lane change or passing maneuver to the right in 103,000 and 14,000 lane change crashes, respectively.
2. Host vehicle passes to the left (longitudinal acceleration less than 0.1g at initial speed of 35-60 mph for light vehicles and 40-55 mph for heavy trucks) and encroaches on an adjacent vehicle in daylight, clear weather, on straight and level road – light vehicle and heavy truck collided with an adjacent vehicle during a lane change or passing maneuver to the left in 108,000 and 6,000 lane change crashes, respectively.
3. Light vehicle turns left at 20-40 mph (heavy truck turns right at 15-35 mph) and encroaches on an adjacent vehicle going straight in daylight, clear weather, on straight and level road – light vehicle collided with an adjacent vehicle during a left turn in 43,000 lane change crashes and heavy truck collided with an adjacent vehicle during a right turn in 10,000 lane change crashes.
4. Host vehicle drifts right (light vehicle at 35-60 mph and heavy truck at 35-55 mph) and encroaches on an adjacent vehicle in daylight, clear weather, on straight and level road – light vehicle and heavy truck collided with an adjacent vehicle while drifting right in 37,000 and 5,000 lane change crashes, respectively.

Run-Off-Road Crash Imminent Test Scenarios

Light vehicles and heavy trucks were reported in 1,028,000 and 41,000 single-vehicle run-off-road crashes. IVBSS target crashes accounted for 85 percent of light vehicle crashes and 67 percent of heavy truck crashes, excluding crashes caused by vehicle failure or evasive maneuver. Five scenarios are recommended as base test scenarios for IVBSS run-off-road countermeasures:

1. Host vehicle is going straight (light vehicle and heavy truck at 25-55 mph) and departs road edge to the right in daylight or darkness, clear weather, on straight and level road – light vehicle and heavy truck departed road edge to the right while going straight in 179,000 and 7,000 run-off-road crashes, respectively.
2. Host vehicle is going straight (light vehicle at 30-60 mph and heavy truck at 25-55 mph) and departs road edge to the left in daylight or darkness, clear weather, on straight and level road – light vehicle and heavy truck departed road edge to the left while going straight in 82,000 and 2,000 run-off-road crashes, respectively.
3. Host vehicle is negotiating a curve (light vehicle and heavy truck at 30-55 mph) and departs road edge to the right in daylight or darkness, clear weather, on sloped road – light vehicle and heavy truck departed road edge to the right while negotiating a curve in 74,000 and 3,000 run-off-road crashes, respectively.
4. Host vehicle is negotiating a curve (light vehicle at 40-60 mph and heavy truck at 35-55 mph) and loses control in daylight, clear or adverse weather, on sloped road – light vehicle and heavy truck lost control while negotiating a curve in 172,000 and 2,000 run-off-road crashes, respectively.
5. Host vehicle is turning left at an intersection (light vehicle at 25-45 mph and heavy truck at 20-40 mph) and departs road edge to the right in daylight, clear weather, on straight and level road – light vehicle and heavy truck departed road edge to the right while initiating a maneuver in 42,000 and 9,000 run-off-road crashes, respectively.

Multiple-Threat Crash Imminent Test Scenarios

This set of crash imminent test scenarios evaluates the capability of the integrated system to issue crash alerts in near simultaneous threat events. There are very few PR crashes in the GES resulting from a prior evasive maneuver to prevent an impending crash. Typically in these cases, the GES does not identify the critical event of the prior evasive maneuver. Thus, the following set of multiple-threat test scenarios is proposed for the light vehicle and heavy truck platforms by combining selected crash imminent test scenarios presented above for rear-end, lane change, and run-off-road crashes:

1. Host vehicle is moving at constant speed and encounters a lead vehicle moving at lower constant speed, host vehicle then attempts to pass to the left adjacent lane occupied by another vehicle.
2. Host vehicle is moving at constant speed and encounters a stopped lead vehicle; host vehicle then attempts to change lanes to the right adjacent lane occupied by another vehicle.
3. Host vehicle drifts and is about to unintentionally depart to the right adjacent lane occupied by another vehicle.
4. Host vehicle drifts and is about to unintentionally depart to the left adjacent lane occupied by another vehicle.
5. Host vehicle is following a lead vehicle at a constant speed on a straight road, both driving too fast for the upcoming curve; and then lead vehicle suddenly decelerates.

1. INTRODUCTION

The U.S. Department of Transportation (USDOT) is working cooperatively with industry partners to accelerate the deployment of integrated rear-end, lane change, and run-off-road crash countermeasure systems under the Integrated Vehicle-Based Safety System (IVBSS) initiative [1]. Such systems will warn drivers when the vehicle they are driving is about to:

- Strike a slower-moving vehicle traveling ahead of them in the same lane and direction of travel;
- Strike a vehicle that is stopped in their travel lane;
- Leave the roadway or traveling too fast for an upcoming curve; or
- Collide with a vehicle in an adjacent lane when the driver is executing a lane change maneuver.

These integrated systems will alert drivers of imminent crash situations to assist them in preventing and reducing the number and severity of injuries resulting from rear-end, lane change, and run-off-road crashes. Preliminary performance guidelines for individual IVBSS functions were provided by USDOT to ensure that system designs meet or exceed the performance capability of systems already evaluated in recent field operational tests [2]. Integration of individual crash countermeasure systems is expected to increase safety benefits, improve overall system performance, reduce system costs, enhance consumer and fleet operator's acceptance, and boost product marketability. The IVBSS initiative is focused on light vehicles (e.g., passenger cars, vans, minivans, sport utility vehicles, and light pickup trucks) and heavy trucks (i.e., gross vehicle weight rating over 10,000 pounds).

The IVBSS initiative will provide USDOT with the information needed to advance the deployment of automotive safety products, including objective test procedures and safety benefits estimation. Objective test procedures promote compliance with performance specifications and allow USDOT to issue consumer information such as safety star ratings. During the execution of the IVBSS initiative, objective test procedures are important to verify that IVBSS prototypes meet their performance specifications and are safe for use by ordinary drivers prior to the start of field operational tests. Typically, objective test procedures consist of a set of crash imminent test scenarios and a set of operational scenarios that are conducted on a closed course [3]. The former set evaluates the capability of a crash warning system to issue timely alerts in crash imminent situations. The latter set assesses the ability of a crash warning system to suppress alerts in conditions that do not pose an immediate threat to the host vehicle.

Crash imminent test scenarios are based on crash data from crash types targeted by IVBSS. Operational scenarios are devised from the capability and limitations of the state-of-the-art technologies in resolving targets and environmental conditions encountered in real-world driving. Objective test procedures will provide a complete set of:

- Initial kinematic conditions for each test scenario;
- Driving conditions;
- Instructions on how to run each scenario;
- Test apparatus (instrumentations and props);

- Measures of performance; and
- Pass-fail criteria.

This report focuses on crash imminent test scenarios and provides a foundation to devise a set of these scenarios based on crash statistics from the 2000-2003 General Estimates System (GES) crash databases [4]. GES obtains its data from a nationally representative probability sample selected from the estimated 6.2 million police-reported crashes that occur annually. In order to calculate estimates of national crash characteristics, data from each police crash report is weighted by a variable called "weight." This variable is the product of the inverse of the probabilities of selection at each of the following three stages in the sampling process: selection of primary sampling units (PSUs); selection of police jurisdictions within a PSU; and selection of crashes for investigation. The national estimates produced from GES data may differ from the true values because they are based on a probability sample of crashes and not a census of all crashes. The size of these differences may vary depending on which sample of crashes was selected [4].

This report identifies most common dynamically distinct pre-crash scenarios leading to rear-end, lane change, and run-off-road crashes for light vehicles and heavy trucks. Pre-crash scenarios describe vehicle movements and critical events immediately prior to the crash [5]. Based on these pre-crash scenarios, this report recommends individual crash imminent test scenarios and statistically describes their environmental factors (weather and lighting conditions), roadway geometry (alignment and profile), and speed conditions (posted speed limit and speeding information). Crash characteristics such as environmental factors and vehicle speed generate test conditions for crash imminent test scenarios. This report also suggests a list of crash imminent test scenarios that examine the capability of the integrated crash warning system in dealing with multiple-threat events. Recommended test scenarios incorporate the dynamic features of all common scenarios and assign a range of test speeds covering the spectrum of all travel speeds reported in crash data. Most severe scenarios are represented in this set of crash imminent test scenarios by capturing most common scenarios in specific crash types and assigning a wide range of test speeds.

Rear-end crashes are defined as the front of a following vehicle striking the rear of a lead vehicle while both are traveling in the same direction [6]. More complicated rear-end crash cases could involve three or more vehicles. The lane change family of crashes typically consists of a crash in which a vehicle attempts to change lanes, merge, pass, leave or enter a parking position, drifts and strikes, or is struck by another vehicle in the adjacent lane, while both are traveling in the same direction. Off-roadway crashes refer to crashes in which the first harmful event occurs off the roadway. Based on GES statistics, there were approximately 3,822,000 police-reported (PR) rear-end, lane change, and run-off-road (ROR) crashes involving all vehicle types in the United States in 2003. These three crash types accounted for about 60 percent of all 6,318,000 PR crashes. There were about 1,775,000 PR rear-end crashes, 570,000 lane change crashes, and 1,478,000 off-roadway crashes involving all vehicle types. These crashes correspond respectively to 28 percent, 9 percent, and 23 percent of all PR crashes in 2003.

2. LIGHT-VEHICLE SCENARIOS

This section describes crash imminent test scenarios for rear-end, lane change, and run-off-road (ROR) crash countermeasure systems on-board light vehicles. Based on GES statistics, light vehicles were involved in approximately 6,060,000 PR crashes or 96 percent of all PR crashes in the United States in 2003. Rear-end, lane change, and ROR crashes accounted for about 3,635,000 PR crashes or 60 percent of all light-vehicle crashes.

2.1. Rear-End Crash Imminent Test Scenarios

Light vehicles were involved in about 1,766,000 PR rear-end crashes based on 2003 GES statistics, accounting for 29 percent of all PR light-vehicle crashes. These crashes involve two or more vehicles per crash. Based on an analysis of two-vehicle crashes, Table 1 identifies the most common pre-crash scenarios that occurred immediately prior to rear-end crashes involving at least one light vehicle. Identification of rear-end crash imminent test scenarios was limited to two-vehicle crashes in order to avoid the complexity of multiple events reported in multi-vehicle (more than two vehicles) crashes. It should be noted that the eight categories listed in Table 1 are mutually exclusive.

As seen in Table 1, two-vehicle rear-end crashes amounted to 1,483,000 PR crashes or 84 percent of all rear-end crashes involving at least one light vehicle. The light vehicle was the striking vehicle in almost 95 percent of these rear-end crashes. The host vehicle will be the striking light vehicle in these scenarios since IVBSS will be designed to assist the driver of the striking vehicle. Scenario 5 is the most frequent scenario, accounting for 27 percent of all two-vehicle rear-end crashes in which the light vehicle is striking. Typically in this scenario, the lead vehicle has just decelerated to a stop and is then struck from behind by a light vehicle. Thus, scenario 5 is considered in this analysis as a lead-vehicle-decelerating scenario. Scenario 6 refers to a lead vehicle that has been stopped for a longer time due to various reasons such as being stuck in stationary traffic. In scenario 1, the lead vehicle is either accelerating, moving at constant speed, decelerating, or stopped in traffic. Appendix A lists the codes used to identify dynamically distinct pre-crash scenarios from GES.

Rear-end pre-crash scenarios highlighted in Table 1are recommended as a basis for the development of crash imminent test scenarios for light-vehicle rear-end crash countermeasures. This basis consists of the following four scenarios:

- A1. Light vehicle makes a maneuver and encounters a lead vehicle (No. 1)
- A2. Light vehicle encounters a lead vehicle moving at a lower constant speed (No. 3)
- A3. Light vehicle encounters a decelerating lead vehicle (No. 4 + No. 5)
- A4. Light vehicle encounters a stopped lead vehicle (No. 6)

In most cases of scenario A1, the following light vehicle is making a lane change and the lead vehicle is stopped. In scenarios A2-A4, the following light vehicle is typically moving at a constant speed.

Table 1. Frequency Distribution of Light-Vehicle Rear-End Pre-Crash Scenarios

No.	Scenario Description	Frequency	Pct.
1	Light vehicle is following and making a maneuver*	75,000	5.1
2	Lead vehicle is accelerating	17,000	1.1
3	Lead vehicle is moving at constant speed	200,000	13.5
4	Lead vehicle is decelerating	347,000	23.4
5	Lead vehicle is stopped in the process of turning or stopped in the presence of a traffic control device	382,000	25.8
6	Lead vehicle is stopped not in the process of turning nor in the presence of traffic control device	364,000	24.6
7	Other rear-end crash scenarios where light vehicle is striking	22,000	1.5
8	Other rear-end crash scenarios where light vehicle is struck	75,000	5.1
	Total	1,482,000	100.0

* Passing, leaving a parked position, entering a parked position, turning right, turning left, making a U-turn, backing up, changing lanes, merging, corrective action, or other.

Appendix B provides GES statistics that describe the driving environment and speed information of scenarios A1-A4. Speed information covers the posted speed limit and whether or not the striking vehicle was speeding at the time of the crash. This speed information substitutes for the actual travel speed of the striking vehicle prior to the crash since over 65 percent of GES cases are coded with unknown travel speed. Table 23 and Table 24 in Appendix B list the statistics of scenarios A1-A4 for light vehicle rear-end crash countermeasures based on two-vehicle rear-end crash data. Key characteristics of these scenarios are:

- Scenarios A1-A4 mainly occur in daylight (greater than or equal to 74%), clear weather (greater than or equal to 84%), and on straight (greater than or equal to 82%) and level (greater than or equal to 76%) roadways.
- The most frequent speed limit is 35 mph in the four scenarios (greater than or equal to 20%).
- More than 20 percent of the crashes in scenarios A1-A4 happen at speed limits less than or equal to 35 mph. At 35 mph speed limit, the striking light vehicle is speeding in over one-third (greater than or equal to 36%) of the crashes in scenarios A2 and A3.
- Over 90 percent of the crashes occur at speed limits less than or equal to 60 mph in scenario A1, less than or equal to 65 mph in scenario A2, and less than or equal to 55 mph in scenarios A3 and A4. The striking light vehicle is speeding in over one-third of the crashes at speed limits 60 mph in scenario A1, 65 mph in scenario A2, and 55 mph in scenario A3.

It is recommended that rear-end crash imminent test scenarios be conducted in daylight, clear weather, and on straight and level roadways. Moreover, these scenarios should be carried out with the striking light vehicle traveling at a low speed and a high speed corresponding respectively to the speed limit associated with more than 20 and 90 percent of the crashes. Selected speed limit should be raised by 10 mph if the striking light vehicle is speeding in over one-third of the crashes at this speed limit. Low travel speeds become 35 mph for scenarios A1 and A4, and 45 mph for scenarios A2 and A3. High travel speeds become 70 mph for scenario A1, 75 mph for scenario A2, 65 mph for scenario A3, and 55 mph for scenario A4. These high speeds could serve as a guide to system design. However, due to safety considerations during the conduct of the objective tests, the speed of the IVBSS-equipped vehicle should not exceed a certain threshold, as determined by professional test track drivers (typically 60 mph).

Table 2 lists the recommended light vehicle, rear-end crash imminent, base test scenarios. It should be noted that scenarios A2 and A3 require information on travel speed of the lead vehicle. In addition, scenario A3 needs information on headway and lead vehicle deceleration. Relative speed values of 5 and 25 mph are recommended for scenario A2. Headway values of 1 second (low speed) and 3 seconds (high speed), and lead vehicle deceleration values of 0.15g (high speed) and 0.35g (low speed), are recommended for scenario A3.

Table 2. Recommended Light-Vehicle, Rear-End Crash Imminent Base Test Scenarios

No.	Crash Imminent Test Scenario
1	Light vehicle changes lanes at 35-60 mph and encounters a stopped lead vehicle in daylight, clear weather, on straight and level road.
2	Light vehicle is moving at constant speed of 45-60 mph and encounters a lead vehicle moving at lower constant speed in daylight, clear weather, on straight and level road.
3	Light vehicle is following a lead vehicle at constant speed of 45-60 mph and then lead vehicle suddenly decelerates in daylight, clear weather, on straight and level road.
4	Light vehicle is moving at constant speed of 35-55 mph and encounters a stopped lead vehicle in daylight, clear weather, on straight and level road.

2.2. Lane Change Crash Imminent Test Scenarios

Light vehicles were involved in about 564,000 PR lane change crashes based on 2003 GES statistics, accounting for 9 percent of all PR light-vehicle crashes. These crashes consist of two or more vehicles per crash. Based on an analysis of two-vehicle crashes, Table 3 identifies the most common pre-crash scenarios that occurred immediately prior to lane change crashes involving at least one light vehicle. Identification of lane change crash imminent test scenarios was limited to two-vehicle crashes in order to avoid the complexity of multiple events reported in multi-vehicle (more than two vehicles) crashes. It should be noted that the 13 categories listed in Table 3 are mutually exclusive.

As seen in Table 3, two-vehicle lane change crashes amounted to 544,000 PR crashes, or 96 percent of all lane change crashes involving at least one light vehicle. The light vehicle encroached into the adjacent lane of other vehicle types in almost 82 percent of these lane change crashes. The host vehicle will be the encroaching light vehicle in these scenarios since IVBSS will be designed to assist the driver of the encroaching vehicle. Scenario 2 is the most frequent scenario (108,000 crashes), accounting for 24 percent of all two-vehicle lane change crashes in which the light vehicle is encroaching onto another vehicle's lane (445,000 crashes). There is a difference between lane change and passing maneuvers in scenarios 1, 2, and 3. In a lane change maneuver, the vehicle changes lanes while maintaining constant longitudinal speed, and the vehicle accelerates while changing lanes during the passing maneuver.

Table 3. Frequency Distribution of Light-Vehicle Lane Change Pre-Crash Scenarios

No.	Scenario Description	Frequency	Pct.
1	Light vehicle changes lanes or passes to the right and encroaches on adjacent vehicle	103,000	19.0
2	Light vehicle changes lanes or passes to the left and encroaches on adjacent vehicle	108,000	19.8
3	Light vehicle is changing lanes or passing to unknown adjacent lane	55,000	10.1
4	Light vehicle merges to the right and encroaches on adjacent vehicle	6,000	1.1
5	Light vehicle merges to the left and encroaches on adjacent vehicle	16,000	3.0
6	Light vehicle is merging to unknown direction	1,000	0.2
7	Light vehicle turns right and encroaches on adjacent vehicle	31,000	5.7
8	Light vehicle turns left and encroaches on adjacent vehicle	43,000	7.9
9	Light vehicle drifts right and encroaches on adjacent vehicle	37,000	6.9
10	Light vehicle drifts left and encroaches on adjacent vehicle	27,000	4.9
11	Light vehicle is encroaching to adjacent lane on the right	7,000	1.3
12	Light vehicle is encroaching to adjacent lane on the left	11,000	1.9
13	Other cases	99,000	18.2
	Total	544,000	100.0

Lane change pre-crash scenarios highlighted in Table 3 are recommended as a basis for the development of crash imminent test scenarios for light vehicle lane change crash countermeasures. There are four scenarios, as follows:

B1. Light vehicle changes lanes or passes to the right and encroaches on adjacent vehicle (No. 1)

B2. Light vehicle changes lanes or passes to the left and encroaches on adjacent vehicle (No. 2)

B3. Light vehicle turns left and encroaches on adjacent vehicle (No. 8)

B4. Light vehicle drifts right and encroaches on adjacent vehicle (No. 9)

Table 4 breaks down scenarios 7 and 8 by the movement of the other vehicle. In 95 percent of the crashes in scenario 7 where the light vehicle was turning right, the other vehicle was going straight. Similarly, the other vehicle was going straight in 91 percent of the crashes in scenario 8 where the light vehicle was turning left.

Table 4. Breakdown of Light-Vehicle Turning Scenarios by Movement of Other Vehicle

Light Vehicle Turning Right			Light Vehicle Turning Left		
Other Vehicle	Frequency	Pct.	Other Vehicle	Frequency	Pct.
Turning right	1,000	2	Turning left	-	1
Going straight	29,000	95	Going straight	39,000	91
Passing	-	0	Passing	3,000	6
Parking	1,000	2	Evasive Maneuver	1,000	2
Changing lanes	-	0	Other	-	1
Other	-	1			
Total	31,000	100	Total	43,000	100

Table 25 and Table 26 in Appendix B show the characteristics of scenarios B1-B4 for light-vehicle lane change crash countermeasures based on two-vehicle crash statistics. Key characteristics are:

- Scenarios B1-B4 mainly occur in daylight (greater than or equal to 67%), clear weather (greater than or equal to 79%), and on straight (greater than or equal to 84%) and level (greater than or equal to 79%) roadways.
- The most frequent speed limit is 35 mph in scenarios B1-B3 and 65 mph in scenario B4.
- More than 20 percent of the crashes happen at speed limits less than or equal to 35 mph in scenarios B1, B2, and B4, and less than or equal to 25 mph in scenario B3. Speeding rates by the encroaching light vehicle are low at these speed limits.
- Over 90 percent of the crashes occur at speed limits less than or equal to 65 mph in scenarios B1, B2, and B4, and less than or equal to 45 mph in scenario B3. At these speed limits, the encroaching light vehicle is speeding in less than one-third of the crashes in each of the four scenarios.

Table 5 lists the recommended light-vehicle, lane change crash imminent, base test scenarios. It should be noted that information is needed on relative speed, relative longitudinal distance, and relative lateral distance between the light vehicle and the other vehicle in the adjacent lane. Relative speeds of 0 and 10 mph, and relative lateral distances of 3 and 9 feet, are recommended at the start of the lane change maneuver. As for relative longitudinal distance, overlapping (side by side) and/or fast approach (other vehicle approaching at a higher speed from a distance) scenarios could be included based on the final design and performance specifications of the system. Moreover, the aggressiveness of the lane change maneuver (e.g., time to change lanes – 2-16 seconds, intended lane change distance – 9-15 feet, and peak lateral acceleration – 0.01g-0.7g [7]) needs to be determined.

Table 5. Recommended Light-Vehicle, Lane Change Crash Imminent Base Test Scenarios

No.	Crash Imminent Test Scenario
1	Light vehicle changes lanes to the right at 35-60 mph and encroaches on adjacent vehicle in daylight, clear weather, on straight and level road. Light vehicle maintains constant longitudinal speed during the lane change maneuver.
2	Light vehicle passes to the left at 35-60 mph and encroaches on adjacent vehicle in daylight, clear weather, on straight and level road. Light vehicle accelerates (approx. 0.1g) during the passing maneuver.
3	Light vehicle turns left at 25-45 mph (or 20-40 mph) and encroaches on adjacent vehicle in daylight, clear weather, on straight and level road.
4	Light vehicle drifts right at 35-60 mph and encroaches on adjacent vehicle in daylight, clear weather, on straight and level road.

2.3. Run-Off-Road Crash Imminent Test Scenarios

Light vehicles were involved in about 1,304,000 PR off-roadway crashes based on 2003 GES statistics, accounting for 22 percent of all PR light-vehicle crashes. These crashes consist of one or more vehicles per crash. Based on an analysis of single-vehicle crashes, Table 6 identifies the most common pre-crash scenarios that occurred immediately prior to a light vehicle running off the road. It should be noted that the ten categories listed in Table 6 are mutually exclusive. As seen in Table 6, single-vehicle run-off-road crashes amounted to 1,028,000 PR crashes or 79 percent of all run-off-road crashes involving at least one light vehicle. Road edge departure and control loss accounted for 43 percent and 42 percent of these crashes, respectively. The light vehicle ran off the right side of the road in two-thirds of road-edge-departure crashes. As indicated in Table 7, this implies that the light vehicle had to cross at least one adjacent travel lane, either same or opposite traffic direction, before departing the road. The light vehicle lost control on a curve in 40 percent of single-vehicle control-loss crashes, mostly due to speeding or prevailing surface conditions. The "other" scenario in Table 6 refers to single-vehicle crashes caused by vehicle failure or evasive maneuver, which are not the target of IVBSS functions.

Table 8 shows the statistics of the first harmful event in all single light-vehicle run-off-road crashes. This event indicates the first property-damaging or injury-producing event in the crash. This information will help identify the type of props that will be placed on the side of the road when conducting run-off-road crash imminent test scenarios. In descending order of frequency, the light vehicle ran into a post, parked vehicle, tree, ditch, or guardrail in 63 percent of run-off-road crashes. Table 35 - Table 38 in Appendix C indicate the most frequent first harmful event in each of the scenarios listed in Table 6.

Table 6. Frequency Distribution of Single Light-Vehicle Run-Off-Road Pre-Crash Scenarios

No.	Scenario Description	Frequency	Pct.
1	Light vehicle is going straight and departs road edge to the right	179,000	17.4
2	Light vehicle is going straight and departs road edge to the left	82,000	8.0
3	Light vehicle is negotiating a curve and departs road edge to the right	74,000	7.2
4	Light vehicle is negotiating a curve and departs road edge to the left	42,000	4.1
5	Light vehicle is going straight and loses control	208,000	20.3
6	Light vehicle is negotiating a curve and loses control	172,000	16.8
7	Light vehicle is initiating a maneuver and loses control	55,000	5.4
8	Light vehicle is initiating a maneuver and departs road edge to the right	42,000	4.1
9	Light vehicle is initiating a maneuver and departs road edge to the left	23,000	2.2
10	Other	150,000	14.6
	Total	1,027,000	100.0

Table 7. Trafficway Flow Information for Light-Vehicle Left-Road-Edge Departure

Trafficway Flow	Scenario 2	Scenario 4
Not Divided	48.9%	66.5%
Divided	32.8%	16.3%
One Way	4.4%	11.1%
Unknown	13.9%	6.1%
Total	100.0%	100.0%

The run-off-road pre-crash scenarios highlighted in Table 6 are recommended as a basis for the development of crash imminent test scenarios for single light-vehicle run-off-road crash countermeasures. There are five scenarios as follows:

C1. Light vehicle is going straight and departs road edge to the right (No. 1)
C2. Light vehicle is going straight and departs road edge to the left (No. 2)
C3. Light vehicle is negotiating a curve and departs road edge to the right (No. 3)
C4. Light vehicle is negotiating a curve and loses control (No. 6)
C5. Light vehicle is initiating a maneuver and departs road edge to the right (No. 8)

In most cases of scenario C5, the light vehicle is turning left at a roadway junction. Table 27 and Table 28 in Appendix B show the characteristics of scenarios C1-C5 for single light-vehicle run-off-road crash countermeasures. Key characteristics are:

- Scenarios C1, C2, and C3 happen at night in 50 percent or more of the crashes.
- Over one-third of the crashes in scenario C4 are reported in adverse weather.
- Scenarios C3 and C4 occurring on curves have higher rates of crashes on sloped roadways (more than 41%) than other scenarios.
- The most frequent speed limit is 25 mph in scenarios C1 and C5, and 55 mph in scenarios C2-C4.
- Over 20 percent of the crashes happen at speed limits less than or equal to 25 mph in scenarios C1 and C5, and less than or equal to 30 mph in scenarios C2-C4. At these two speed limits, a high speeding rate is only observed in scenario C4 among the five scenarios.
- Over 90 percent of the crashes occur at speed limits less than or equal to 55 mph in scenario C1 and scenarios C3-C5, and less than or equal to 65 mph in C2. The light vehicle is speeding in over two-thirds of the crashes in scenario C4 at 55 mph speed limit.

Table 9 lists the recommended single light-vehicle, run-off-road crash imminent, base test scenarios. In scenario 4, IVBSS will be designed to alert the driver about excessive speed for an upcoming curve before the vehicle enters the curve and loses control. It should be noted that curve radius and elevation data are needed for scenarios C3 and C4. Vehicle departure angles, as well as road markings, are also required.

Table 8. First Harmful Event Statistics of Single Light-Vehicle Run-Off-Road Crashes

First Harmful Event	Frequency	Pct.
Post, pole, or support	173,000	16.9
Parked vehicle	165,000	16.1
Tree	110,000	10.7
Culvert or ditch	105,000	10.2
Guardrail	90,000	8.7
Rollover/overturn	71,000	6.9
Traffic barrier	58,000	5.7
Embankment	46,000	4.5
Curb	41,000	4.0
Other object not fixed	39,000	3.8
Fence	35,000	3.4
Other fixed object	31,000	3.0
Bridge structure	16,000	1.5
Wall	9,000	0.9
Fire hydrant	7,000	0.7
Building	6,000	0.6
Boulder	6,000	0.6
Shrubbery or bush	6,000	0.5
Fixed object - no details	3,000	0.3
Ground	2,000	0.2
Impact attenuator	2,000	0.2
Pavement surface irregularity	2,000	0.2
Railway train	2,000	0.2
Thrown or falling object	1,000	0.1
Object not fixed - no details	1,000	0.1
Pedestrian	1,000	0.1
Animal	-	0.03
Other type non-motorist	-	0.03
Other noncollision	-	0.02
Cyclist	-	0.003
Total	1,028,000	100.0

Table 9. Recommended Light-Vehicle, Run-Off-Road Crash Imminent Base Test Scenarios

No.	Crash Imminent Test Scenario
1	Light vehicle is going straight at 25-55 mph and departs road edge to the right in daylight or darkness,* clear weather, on straight and level road.
2	Light vehicle is going straight at 30-60 mph and departs road edge to the left in daylight or darkness,* clear weather, on straight and level road.
3	Light vehicle is negotiating a curve at 30-55 mph and departs road edge to the right in daylight or darkness,* clear weather, on a sloping road.
4	Light vehicle is negotiating a curve at 40-60 mph and loses control in daylight, clear or adverse** (slippery surface) weather, on a sloping road.
5	Light vehicle is turning left at 25-45 mph and departs road edge to the right in daylight, clear weather, on straight and level road.

* Test should be conducted in daylight and repeated in darkness
** Test should be conducted on a dry surface and repeated on a slippery surface

3. HEAVY-TRUCK SCENARIOS

This section describes crash imminent test scenarios for rear-end, lane change, and run-off-road crash countermeasure systems on-board heavy trucks. Based on GES statistics, heavy trucks were involved in about 362,000 PR crashes or 6 percent of all PR crashes in the United States in 2003. Rear-end, lane change, and off-roadway crashes accounted for about 216,000 PR crashes or 60 percent of all heavy-truck crashes.

3.1. Rear-End Crash Imminent Test Scenarios

Heavy trucks were involved in about 78,000 PR rear-end crashes based on 2003 GES statistics, accounting for 22 percent of all PR heavy-truck crashes. These crashes consist of two or more vehicles per crash. Identification of rear-end crash imminent test scenarios was limited to two-vehicle crashes in order to avoid the complexity of multiple events reported in multi-vehicle (more than two vehicles) crashes. Heavy truck involvement in two-vehicle rear-end crashes was estimated at 64,000 PR crashes or 82 percent of all heavy-truck rear-end crashes in 2003. Based on an analysis of two-vehicle crashes in 2000-2003 GES databases, Table 10 lists the most common pre-crash scenarios that occurred immediately prior to rear-end crashes involving at least one heavy truck. Four years of GES data were used to identify pre-crash scenarios for heavy trucks due to the low number of crash cases representing heavy trucks each year. It should be noted that the eight categories listed in Table 10 are mutually exclusive.

The heavy truck was the striking vehicle in about 60 percent of these rear-end crashes. The host vehicle will be the striking heavy truck in these scenarios, since IVBSS will be designed to assist the driver of the striking vehicle. Scenario 6 is the most frequent scenario (49,000 crashes), accounting for 32 percent of all two-vehicle rear-end crashes in which the heavy truck was striking (154,000 crashes). Scenario 5 is considered in this analysis as a lead vehicle-decelerating scenario because, typically, the lead vehicle had just decelerated to a stop before being struck by the heavy truck.

Table 10. Frequency Distribution of Heavy Truck Rear-End Pre-Crash Scenarios

No.	Scenario Description	Frequency	Pct.
1	Heavy truck is following and making a maneuver*	10,000	3.9
2	Lead vehicle is accelerating	2,000	0.7
3	Lead vehicle is moving at constant speed	34,000	13.3
4	Lead vehicle is decelerating	37,000	14.5
5	Lead vehicle is stopped in the process of turning or stopped in the presence of a traffic control device	17,000	6.5
6	Lead vehicle is stopped not in the process of turning nor in the presence of traffic control device	49,000	19.5
7	Other rear-end crash scenarios where heavy truck is striking	5,000	1.9
8	Other rear-end crash scenarios where heavy truck is struck	100,000	39.6
	Total	254,000	100.0

* Passing, leaving a parked position, entering a parked position, turning right, turning left, making a U-turn, backing up, changing lanes, merging, corrective action, or other.

Rear-end pre-crash scenarios highlighted in Table 10 are recommended as a basis for the development of crash imminent test scenarios for the following heavy truck rear-end crash countermeasures:

- D1. Heavy truck makes a maneuver and encounters a stopped lead vehicle (No. 1)
- D2. Heavy truck encounters a lead vehicle moving at lower constant speed (No. 3)
- D3. Heavy truck encounters a decelerating lead vehicle (No. 4 + No. 5)
- D4. Heavy truck encounters a stopped lead vehicle (No. 6)

In most cases of scenario D1, the following heavy truck is making a lane change and the lead vehicle is stopped. In scenarios D2-D4, the following heavy truck is typically moving at a constant speed. Table 29 and Table 30 in Appendix B show the characteristics of scenarios D1-D4 for heavy truck rear-end crash countermeasures based on two-vehicle crash statistics. Key characteristics are:

- Scenarios D1-D4 mainly occur in daylight (greater than or equal to 71%), clear weather (greater than or equal to 86%), and on straight (greater than or equal to 89%) and level (greater than or equal to 73%) roadways.
- The most frequent speed limit is 45 mph in scenario D1, 55 mph in scenarios D2 and D3, and 35 mph in scenario D4.
- Over 20 percent of the crashes happen at speed limits less than or equal to 35 mph in scenarios D1, D3, and D4, and less than or equal to 45 mph in scenario D2. Low speeding rates are observed at these speed limits.
- Over 90 percent of the crashes occur at speed limits less than or equal to 65 mph in scenarios D1 and D3, less than or equal to 75 mph in scenario D2, and less than or equal to 55 mph in scenario D4. At 65 mph speed limit, the heavy truck is speeding in 33 percent of the crashes in scenario D3. The heavy truck is also speeding in 41 percent of the crashes in scenario D2 at speed limit greater than or equal to 70 mph. The reader is cautioned about these statistics at very high speed limits due to the small number of cases representing these crashes.

Due to safety considerations during the conduct of heavy truck objective tests, the speed of heavy trucks should not exceed a certain threshold, as determined by professional test track drivers (typically 55 mph). Table 11 lists the recommended heavy truck, rear-end crash imminent, base test scenarios.

Table 11. Recommended Heavy Truck, Rear-End Crash Imminent Base Test Scenarios

No.	Crash Imminent Test Scenario
1	Heavy truck changes lanes at 35-55 mph and encounters a stopped lead vehicle in daylight, clear weather, on straight and level road.
2	Heavy truck is moving at constant speed of 45-55 mph and encounters a lead vehicle moving at lower constant speed in daylight, clear weather, on straight and level road.
3	Heavy truck is following a lead vehicle at constant speed of 35-55 mph and then lead vehicle suddenly decelerates in daylight, clear weather, on straight and level road.
4	Heavy truck is moving at constant speed of 35-55 mph and encounters a stopped lead vehicle in daylight, clear weather, on straight and level road.

3.2. Lane Change Crash Imminent Test Scenarios

Heavy trucks were involved in about 82,000 PR lane change crashes based on 2003 GES statistics, accounting for 23 percent of all PR heavy-truck crashes. These crashes involve two or more vehicles per crash. Identification of lane change crash imminent test scenarios was limited to two-vehicle crashes in order to avoid the complexity of multiple events reported in multi-vehicle (more than two vehicles) crashes. Heavy truck involvement in two-vehicle lane change crashes was estimated at 76,000 PR crashes or 93 percent of all heavy-truck lane change crashes in 2003. Based on an analysis of two-vehicle crashes in 2000-2003 GES databases, Table 12 identifies the most common pre-crash scenarios that occurred immediately prior to lane change crashes involving at least one heavy truck. It should be noted that the 13 categories listed in Table 12 are mutually exclusive.

The heavy truck encroached into the adjacent lane of other vehicle types in almost 59 percent of these lane change crashes. The host vehicle will be the encroaching heavy truck in these scenarios since IVBSS will be designed to assist the driver of the encroaching vehicle. Scenario 1 is the most frequent scenario (58,000 crashes), accounting for 31 percent of all two-vehicle lane change crashes in which the heavy truck was encroaching onto another vehicle's lane (184,000 crashes).

Table 13 breaks down scenarios 7 and 8 by the movement of the other vehicle. In 52 percent of the crashes in scenario 7 where the heavy truck was turning right, the other vehicle was going straight. The other vehicle was either passing the heavy truck or stopped in 23 and 21 percent of the crashes in this same scenario, respectively. The other vehicle was passing the heavy truck in 49 percent of the crashes in scenario 8 where the heavy truck was turning left. In two-vehicle turning crashes, heavy trucks were turning right 2.5 times more than turning left. By contrast, light vehicles were turning left 1.4 times more than turning right in two-vehicle turning crashes.

Table 12. Frequency Distribution of Heavy-Truck Lane Change Pre-Crash Scenarios

No.	Scenario Description	Frequency	Pct.
1	Heavy truck changes lanes or passes to the right and encroaches on adjacent vehicle	58,000	18.5
2	Heavy truck changes lanes or passes to the left and encroaches on adjacent vehicle	25,000	8.1
3	Heavy truck is changing lanes or passing to unknown adjacent lane	5,000	1.6
4	Heavy truck merges to the right and encroaches on adjacent vehicle	1,000	0.4
5	Heavy truck merges to the left and encroaches on adjacent vehicle	5,000	1.7
6	Heavy truck is merging to unknown direction	1,000	0.3
7	Heavy truck turns right and encroaches on adjacent vehicle	38,000	12.2
8	Heavy truck turns left and encroaches on adjacent vehicle	15,000	4.9
9	Heavy truck drifts right and encroaches on adjacent vehicle	18,000	5.8
10	Heavy truck drifts left and encroaches on adjacent vehicle	16,000	5.0
11	Heavy truck is encroaching to adjacent lane on the right	1,000	0.5
12	Heavy truck is encroaching to adjacent lane on the left	1,000	0.3
13	Other	127,000	40.7
	Total	311,000	100.0

Table 13. Breakdown of Heavy-Truck Turning Scenarios by Movement of Other Vehicle

Heavy Truck Turning Right			Heavy Truck Turning Left		
Other Vehicle	Frequency	Pct.	Other Vehicle	Frequency	Pct.
Going straight	20,000	52	Going straight	5,000	36
Stopped	8,000	21	Stopped	2,000	12
Passing	9,000	23	Passing	7,000	49
Parking	1,000	1	Turning left	-	0
Turning right	-	0	Other	-	3
Changing lanes	-	0			
Other	1,000	2			
Total	38,000	100	Total	15,000	100

Lane change pre-crash scenarios highlighted in Table 12 are recommended as a basis for the development of crash imminent test scenarios for heavy-truck lane change crash countermeasures. The four scenarios are as follows:

E1. Heavy truck changes lanes or passes to the right and encroaches on adjacent vehicle (No. 1)
E2. Heavy truck changes lanes or passes to the left and encroaches on adjacent vehicle (No. 2)
E3. Heavy truck turns right and encroaches on adjacent vehicle (No. 7)
E4. Heavy truck drifts right and encroaches on adjacent vehicle (No. 9)

Table 31 and Table 32 in Appendix B show the characteristics of scenarios E1-E4 for heavy truck lane change crash countermeasures based on two-vehicle crash statistics. Key characteristics are:

- Scenarios E1-E4 mainly occur in daylight (greater than or equal to 73%), clear weather (greater than or equal to 88%), and on straight (greater than or equal to 88%) and level (greater than or equal to 73%) roadways.
- The most frequent speed limit is 55 mph in scenarios E1-E2 and 35 mph in scenarios E3-E4.
- Over 20 percent of the crashes happen at speed limits less than or equal to 35 mph in scenarios E1 and E4, less than or equal to 40 mph in scenario E2, and less than or equal to 25 mph in scenario E3. Speeding rates by the encroaching heavy truck are very low at these speed limits.
- Over 90 percent of the crashes occur at speed limits less than or equal to 65 mph in scenarios E1 and E4, less than or equal to 75 mph in scenario E2, and less than or equal to 55 mph in scenario E3. At these speed limits, the encroaching heavy truck is not speeding in scenarios E1-E3. At a 65 mph speed limit, the heavy truck is speeding in 20 percent of the crashes in scenario E4.

Table 14 lists the recommended base test scenarios for heavy trucks with a lane change crash imminent.

Table 14. Recommended Heavy-Truck, Lane Change Crash Imminent Base Test Scenarios

No.	Crash Imminent Test Scenario
1	Heavy truck changes lanes to the right at 35-55 mph and encroaches on adjacent vehicle in daylight, clear weather, on straight and level road. Heavy truck maintains constant longitudinal speed during the lane change maneuver.
2	Heavy truck passes to the left at 40-55 mph and encroaches on adjacent vehicle in daylight, clear weather, on straight and level road. Heavy truck accelerates during the passing maneuver.
3	Heavy truck turns right at 25-45 mph (or 15-35 mph) and encroaches on adjacent vehicle in daylight, clear weather, on straight and level road.
4	Heavy truck drifts right at 35-55 mph and encroaches on adjacent vehicle in daylight, clear weather, on straight and level road.

3.3. Run-Off-Road Crash Imminent Test Scenarios

Heavy trucks were involved in about 55,000 PR off-roadway crashes based on 2003 GES statistics, accounting for 15 percent of all PR heavy-truck crashes. These crashes consist of one or more vehicles per crash. Heavy-truck involvement in single-vehicle run-off-road crashes was estimated at 38,000 PR crashes or 69 percent of all run-off-road crashes involving at least one heavy truck.

Road edge departure and control loss accounted for 54 percent and 13 percent of these crashes, respectively. The heavy truck ran off the right side of the road in 83 percent of road edge departure crashes. As indicated in Table 16 regarding left road edge departure, the trafficway was not divided in only 40 percent and 36 percent respectively of scenario 2 and scenario 4 crashes. Unlike light vehicle crash test scenarios, it is not recommended that left-edge-departure tests for heavy trucks include an additional adjacent lane of travel (crossing an adjacent lane of travel before departing the road edge). The heavy truck lost control on a curve in 43 percent of single-vehicle control-loss crashes. The "other" scenario in Table 15 lists the most common pre-crash scenarios that occurred immediately prior to run-off-road crashes involving heavy trucks based on 2000-2003 GES statistics. Table 15 refers to single-vehicle crashes caused by vehicle failure or evasive maneuver, which are not the target of IVBSS functions. It should be noted that the ten categories listed in Table 15 are mutually exclusive.

Table 15. Frequency Distribution of Heavy-Truck Run-Off-Road Pre-Crash Scenarios

No.	Scenario Description	Frequency	Pct.
1	Heavy truck is going straight and departs road edge to the right	28,000	16.9
2	Heavy truck is going straight and departs road edge to the left	7,000	4.5
3	Heavy truck is negotiating a curve and departs road edge to the right	11,000	6.5
4	Heavy truck is negotiating a curve and departs road edge to the left	2,000	1.3
5	Heavy truck is going straight and loses control	7,000	4.3
6	Heavy truck is negotiating a curve and loses control	9,000	5.6
7	Heavy truck is initiating a maneuver and loses control	5,000	3.1
8	Heavy truck is initiating a maneuver and departs road edge to the right	36,000	21.5
9	Heavy truck is initiating a maneuver and departs road edge to the left	5,000	3.3
10	Other	55,000	33.2
	Total	165,000	100.0

Table 16. Trafficway Flow Information for Heavy-Truck Left-Road-Edge Departure

Trafficway Flow	Scenario 2	Scenario 4
Not Divided	40%	36%
Divided	44%	22%
One Way	2%	40%
Unknown	14%	2%
Total	100%	100%

Table 17 presents the statistics of the first harmful event in all heavy truck run-off-road crashes. In descending order of frequency, the heavy truck ran into a parked vehicle, hit a post, rolled over, hit other fixed object, or hit a guardrail in 71 percent of these crashes. Table 39 - Table 42 in Appendix C indicate the most frequent first harmful event in each of the scenarios listed in Table 15 lists the most common pre-crash scenarios that occurred immediately prior to run-off-road crashes involving heavy trucks based on 2000-2003 GES statistics. It should be noted that the ten categories listed in Table 15 are mutually exclusive. The reader is cautioned that heavy-truck crash statistics in this report were drawn from only four years of GES data; thus, these statistics are based on weighted data from relatively few cases of heavy-truck crashes.

Table 17. First Harmful Event Statistics of Heavy-Truck Run-Off-Road Crashes

First Harmful Event	Frequency	Pct.
Parked Motor Vehicle	47,000	28.5
Post, Pole or Support	26,000	15.5
Rollover/Overturn	16,000	9.5
Other Fixed Object	16,000	9.5
Guardrail	13,000	7.7
Other Object Not Fixed	10,000	5.9
Culvert or Ditch	8,000	5.1
Tree	4,000	2.6
Bridge Structure	4,000	2.6
Fence	4,000	2.2
Embankment	4,000	2.2
Traffic Barrier	3,000	1.6
Fire Hydrant	2,000	1.4
Curb	2,000	1.4
Building	1,000	0.9
Railway Train	1,000	0.7
Ground	1,000	0.7
Shrubbery or Bush	1,000	0.6
Wall	1,000	0.5
Object Not Fixed-No Details	1,000	0.4
Impact Attenuator	1,000	0.3
Fixed Object, No Details	-	0.2
Boulder	-	0.2
Thrown or Falling Object	-	0.02
Pedestrian	-	0.01
Pavement Irregularity	-	0.001
Total	166,000	100.0

Run-off-road pre-crash scenarios highlighted in Table 15 are recommended as a basis for the development of crash imminent test scenarios for heavy-truck run-off-road crash countermeasures. There are five base scenarios, as follows:

F1. Heavy truck is going straight and departs road edge to the right (No. 1)
F2. Heavy truck is going straight and departs road edge to the left (No. 2)
F3. Heavy truck is negotiating a curve and departs road edge to the right (No. 3)
F4. Heavy truck is negotiating a curve and loses control (No. 6)
F5. Heavy truck is initiating a maneuver and departs road edge to the right (No. 8)

In most cases of scenario F5, the heavy truck is turning left at a roadway junction. Table 33 and Table 34 in Appendix B show the characteristics of scenarios F1-F5 for heavy-truck run-off-road crash countermeasures. Key characteristics are:

- Scenarios F1 and F3-F5 occur mostly in daylight (greater than or equal to 71%), while scenario F2 has the highest frequency at night with 61 percent of the crashes.
- Scenarios F1-F5 happen in clear weather (greater than or equal to 80%).
- Scenarios F3 and F4 occurring on curves have higher rates of crashes on sloped roadways (greater than 53%) than other scenarios.
- The most frequent speed limit is less than or equal to 25 mph in scenarios F1, F2, and F5, and 55 mph in scenarios F3 and F4.
- Over 20 percent of the crashes happen at speed limits less than or equal to 25 mph in scenarios F1, F2, and F5, less than or equal to 30 mph in scenario F3, and less than or equal to 35 mph in scenario F4. At these speed limits, a high speeding rate is only observed in scenario F4 among the five scenarios.
- Over 90 percent of the crashes occur at speed limits less than or equal to 75 mph in scenarios F1 and F2, less than or equal to 55 mph in scenario F3, less than or equal to 65 mph in scenario F4, and less than or equal to 45 mph in scenario F5. The heavy truck is speeding in over 60 percent of the crashes in scenario F2 at greater than or equal to 70 mph speed limit, and in over three-quarters of the crashes in scenario F4 at 65 mph speed limit. The reader is cautioned about these speeding statistics at very high speed limits due to the small number of cases representing these crashes.

Table 18 lists the recommended heavy-truck, run-off-road crash imminent, base test scenarios. In scenario 4, IVBSS will be designed to alert the driver about excessive speed for an upcoming curve before the vehicle enters the curve and loses control.

Table 18. Recommended Heavy Truck, Run-Off-Road Crash Imminent Base Test Scenarios

No.	Crash Imminent Test Scenario
1	Heavy truck is going straight at 25-55 mph and departs road edge to the right in daylight, clear weather, on straight and level road.
2	Heavy truck is going straight at 25-55 mph and departs road edge to the left in daylight or darkness,* clear weather, on straight and level road.
3	Heavy truck is negotiating a curve at 30-55 mph and departs road edge to the right in daylight, clear weather, on a sloping road.
4	Heavy truck is negotiating a curve at 35-55 mph and loses control in daylight, clear weather, on sloping road.
5	Heavy truck is turning left at 20-40 mph and departs road edge to the right in daylight, clear weather, on straight and level road.

* Test should be conducted in daylight and repeated in darkness

4. MULTIPLE-THREAT CRASH IMMINENT TEST SCENARIOS

This set of crash imminent test scenarios evaluates the capability of the integrated system to issue crash alerts in near simultaneous threat events. There are very few PR crashes in GES that involve one vehicle taking a prior evasive maneuver to prevent a crash and then being involved in another crash. Typically in these cases, GES does not identify the critical event associated with the prior evasive maneuver. Thus, the following set of multiple-threat test scenarios is proposed for the light-vehicle and heavy-truck platforms by combining selected scenarios presented earlier in this report.

4.1. Rear-End and Lane Change Crash Imminent Test Scenarios

Two scenarios are suggested by combining scenarios A2/D2 and A4/D4 from rear-end crash imminent test scenarios and scenarios B1/E1 and B2/E2 from lane change crash imminent test scenarios:

1. Host vehicle is moving at a constant speed and encounters a lead vehicle moving at a lower constant speed (rear-end, A2/D2), the host vehicle then attempts to pass to the left adjacent lane that is occupied by another vehicle (lane change, B2/E2). It is recommended that the test be conducted so that the integrated system issues a lane change crash warning first by passing the slower vehicle, getting the alert from the adjacent vehicle, and then returning to the original lane to trigger an alert from the slower vehicle ahead. This test could be repeated so that the integrated system issues a rear-end crash warning first by approaching the slower vehicle to trigger the alert and then passing to obtain the alert from the adjacent vehicle.

2. Host vehicle is moving at a constant speed and encounters a stopped lead vehicle (rear-end, A4/D4), the host vehicle then attempts to change lanes to the right adjacent lane that is occupied by another vehicle (lane change, B1/E1). It is recommended that this test be conducted as suggested above for the first multiple-threat scenario.

4.2. Lane Change and Run-Off-Road Crash Imminent Test Scenarios

Two scenarios are suggested by combining scenario B4/E4 and scenario 10 from lane change crashes and scenarios C1/F1 and C2/F2 from run-off-road crash imminent test scenarios:

1. Host vehicle is going straight, drifts, and is about to depart to the right (run-off-road, C1/F1) adjacent lane that is occupied by another vehicle (lane change, B4/E4). This is considered an unintended lane departure if the right turn signal is not activated. It is recommended that the test be conducted so that the integrated system issues a lane change crash warning first, and then be repeated so that the integrated system issues a run-off-road crash or lane departure warning first (the feasibility of this part of the test, however, may depend on system design).

2. Host vehicle is going straight, drifts, and is about to depart to the left (run-off-road, C2/F2) adjacent lane that is occupied by another vehicle (lane change, 10). This is

considered an unintended lane departure if the left turn signal is not activated. It is recommended that this test be conducted as suggested above for the first multiple-threat scenario.

4.3. Rear-End and Run-Off-Road Crash Imminent Test Scenarios

The following scenario combines scenario A3/D3 from rear-end crash imminent test scenarios and scenario C4/F4 from run-off-road crash imminent test scenarios:

1. Host vehicle is moving at a constant speed on a straight road and encounters a decelerating lead vehicle (rear-end, A3/D3), while driving too fast for the upcoming curve (run-off-road, C4/F4). It is recommended that the test be conducted so that the integrated system issues a rear-end crash warning first, and then be repeated so that the integrated system issues a run-off-road crash (curve speed) warning first.

5. RECOMMENDED CRASH IMMINENT BASE TEST SCENARIOS

Table 19 - Table 22 summarize the results of this study by listing the recommended crash imminent test scenarios for the light-vehicle and heavy-truck platforms respectively for rear-end, lane change, run-off-road, and multiple-threat crash countermeasure functions.

Table 19. Recommended Rear-End Crash Imminent Base Test Scenarios for Light Vehicle and Heavy Truck

No.	Light Vehicle (Host Vehicle)	Heavy Truck (Host Vehicle)	Schematic
1	Light vehicle changes lanes at 35-60 mph and encounters a stopped lead vehicle in daylight, clear weather, on straight and level road.	Heavy truck changes lanes at 35-55 mph and encounters a stopped lead vehicle in daylight, clear weather, on straight and level road.	
2	Light vehicle is moving at constant speed of 45-60 mph and encounters a lead vehicle moving at lower constant speed in daylight, clear weather, on straight and level road.	Heavy truck is moving at constant speed of 45-55 mph and encounters a lead vehicle moving at lower constant speed in daylight, clear weather, on straight and level road.	
3	Light vehicle is following a lead vehicle at constant speed of 45-60 mph and then lead vehicle suddenly decelerates in daylight, clear weather, on straight and level road.	Heavy truck is following a lead vehicle at constant speed of 35-55 mph and then lead vehicle suddenly decelerates in daylight, clear weather, on straight and level road.	
4	Light vehicle is moving at constant speed of 35-55 mph and encounters a stopped lead vehicle in daylight, clear weather, on straight and level road.	Heavy truck is moving at constant speed of 35-55 mph and encounters a stopped lead vehicle in daylight, clear weather, on straight and level road.	

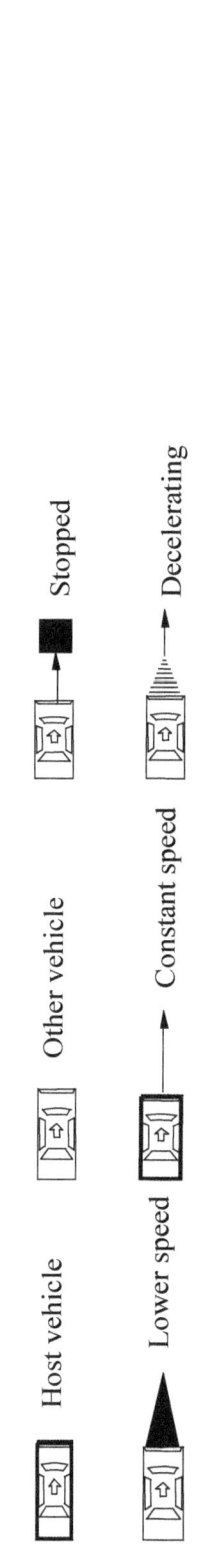

Host vehicle   Other vehicle   Constant speed   Stopped

Lower speed   Decelerating

Changing lanes

22

Table 20. Recommended Lane Change Crash Imminent Base Test Scenarios for Light Vehicles and Heavy Trucks

No.	Light Vehicle (Host Vehicle)	Heavy Truck (Host Vehicle)	Schematic
1	Light vehicle changes lanes to the right at 35-60 mph and encroaches on adjacent vehicle in daylight, clear weather, on straight and level road. Light vehicle maintains constant longitudinal speed during the lane change maneuver.	Heavy truck changes lanes to the right at 35-55 mph and encroaches on adjacent vehicle in daylight, clear weather, on straight and level road. Heavy truck maintains constant longitudinal speed during the lane change maneuver.	
2	Light vehicle passes to the left at 35-60 mph and encroaches on adjacent vehicle in daylight, clear weather, on straight and level road. Light vehicle accelerates (approx. 0.1g) during the passing maneuver.	Heavy truck passes to the left at 40-55 mph and encroaches on adjacent vehicle in daylight, clear weather, on straight and level road. Heavy truck accelerates during the passing maneuver.	
3	Light vehicle turns left at 25-45 mph (or 20-40 mph) and encroaches on adjacent vehicle in daylight, clear weather, on straight and level road.	Heavy truck turns left at 25-45 mph (or 15-35 mph) and encroaches on adjacent vehicle in daylight, clear weather, on straight and level road.	
4	Light vehicle drifts right at 35-60 mph and encroaches on adjacent vehicle in daylight, clear weather, on straight and level road.	Heavy truck drifts right at 35-55 mph and encroaches on adjacent vehicle in daylight, clear weather, on straight and level road.	

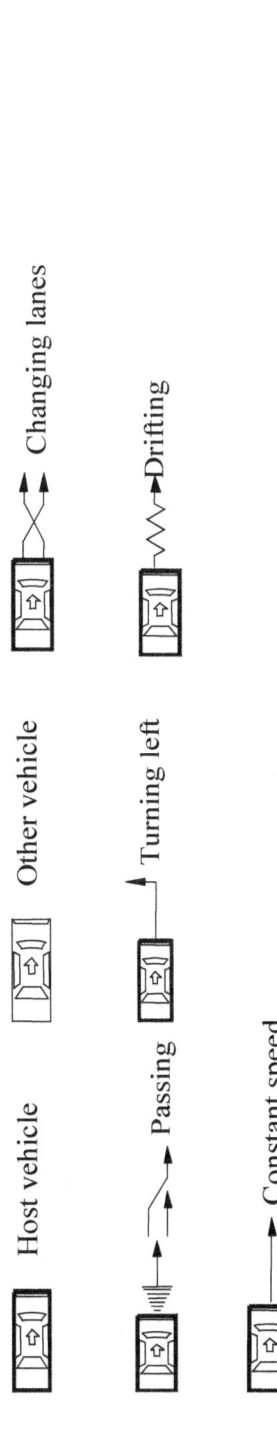

Host vehicle   Other vehicle   Changing lanes

Passing   Turning left   Drifting

Constant speed

Table 21. Recommended Run-Off-Road Crash Imminent Base Test Scenarios for Light Vehicles and Heavy Trucks

No.	Light Vehicle (Host Vehicle)	Heavy Truck (Host Vehicle)	Schematic
1	Light vehicle is going straight at 25-55 mph and departs road edge to the right in daylight or darkness, clear weather, on straight and level road.	Heavy truck is going straight at 25-55 mph and departs road edge to the right in daylight, clear weather, on straight and level road.	
2	Light vehicle is going straight at 30-60 mph and departs road edge to the left in daylight or darkness, clear weather, on straight and level road.	Heavy truck is going straight at 25-55 mph and departs road edge to the left in daylight/ darkness, clear weather, on straight and level road.	
3	Light vehicle is negotiating a curve at 30-55 mph and departs road edge to the right in daylight or darkness, clear weather, on sloping road.	Heavy truck is negotiating a curve at 30-55 mph and departs road edge to the right in daylight, clear weather, on sloping road.	
4	Light vehicle is negotiating a curve at 40-60 mph and loses control in daylight, clear or adverse (i.e., slippery surface) weather, on sloping road.	Heavy truck is negotiating a curve at 35-55 mph and loses control in daylight, clear weather, on sloping road.	
5	Light vehicle is turning left at 25-45 mph and departs road edge to the right in daylight, clear weather, on straight and level road.	Heavy truck is turning left at 20-40 mph and departs road edge to the right in daylight, clear weather, on straight and level road.	

Departing right road edge

Departing left road edge

Control loss

Table 22. Recommended Multiple-Threat Base Test Scenarios for Light Vehicles and Heavy Trucks

No	Light Vehicle and Heavy Truck (Host Vehicles)	Schematic
1	Host vehicle is moving at constant speed and encounters a lead vehicle moving at lower constant speed, host vehicle then attempts to pass to the left adjacent lane that is occupied by another vehicle.	
2	Host vehicle is moving at constant speed and encounters a stopped lead vehicle, host vehicle then attempts to change lanes to the right adjacent lane that is occupied by another vehicle.	
3	Host vehicle drifts and is about to unintentionally depart to the right adjacent lane (unintentional lane departure threat) that is occupied by another vehicle (lane change crash threat).	
4	Host vehicle drifts and is about to unintentionally depart to the left adjacent lane (unintentional lane departure threat) that is occupied by another vehicle (lane change crash threat).	
5	Host vehicle is following a lead vehicle at same constant speed on a straight road, both driving too fast for the upcoming curve. Lead vehicle suddenly decelerates.	

 Speeding

6. CONCLUSIONS

This report recommended a basic set of crash imminent test scenarios based on most common pre-crash scenarios for integrated vehicle based safety systems that alert the driver of a light vehicle or a heavy truck to an impending rear-end, lane change, or run-off-road crash. This report also suggested a list of crash imminent test scenarios that examine the capability of the integrated crash warning system in dealing with multiple-threat scenarios. Pre-crash scenarios describe vehicle movements and critical events immediately prior to the crash. The GES crash database was queried to distinguish most common pre-crash scenarios for light vehicles (2003 GES) and heavy trucks (2000-2003 GES) in terms of their frequency of occurrence. The report also statistically described individual scenarios in terms of their environmental factors (weather and lighting conditions), roadway geometry (alignment and profile), and speed conditions (posted speed limit and speeding information).

Analysis of two-vehicle rear-end crashes revealed four dominant scenarios that accounted for 97 percent of light-vehicle rear-end crashes and 95 percent of heavy truck rear-end crashes in which the subject vehicle was striking. Four scenarios were also identified from an analysis of two-vehicle lane change crashes, comprising 65 percent of light-vehicle crashes and 76 percent of heavy-truck crashes in which the subject vehicle was encroaching onto another vehicle in adjacent lanes. There were five single-vehicle run-off-road scenarios representing 63 percent of light-vehicle crashes and 83 percent of heavy-truck crashes, excluding crashes caused by vehicle failure or evasive maneuver. Based on selected combinations of these individual scenarios, five additional test scenarios were proposed to address multiple threats from near simultaneous critical events.

Further analysis is recommended to describe target pre-crash scenarios in terms of their injury severity so as to ensure that proposed crash imminent test scenarios represent most severe scenarios as well as most frequent scenarios. This analysis will identify the property-damage-only vehicles and will examine the maximum injury severity suffered by each person involved in the crash using the 2003 GES for light vehicles and the 2000-2003 GES for heavy trucks.

7. REFERENCES

[1] Ference, J.J., *The Integrated Vehicle-Based Safety Systems Initiative*. Proceedings of the ITS World Congress, London, UK, October 8-12, 2006.

[2] National Highway Traffic Safety Administration, *Discretionary Cooperative Agreement for Integrated Vehicle-Based Safety Systems (IVBSS) – Request for Applications*. Appendix A, U. S. Department of Transportation, National Highway Traffic Safety Administration, Washington, DC, 2005.

[3] Ference, J.J., Szabo, S., and Najm, W.G., *Performance Evaluation of Integrated Vehicle-Based Safety Systems*. Proceedings of the Performance Metrics for Intelligent Systems (PerMIS) Workshop, National Institute of Standards and Technology, Gaithersburg, MD, August 21 - 23, 2006.

[4] National Center for Statistics and Analysis, *National Automotive Sampling System (NASS) General Estimates System (GES) Analytical User's Manual 1988-2004*. U. S. Department of Transportation, National Highway Traffic Safety Administration, Washington, DC, 2005.

[5] Najm, W.G., Koopmann, J.A., Boyle, L. Ng, and Smith, D.L., *Development of Test Scenarios for Off-Roadway Crash Countermeasures Based on Crash Statistics*. U. S. Department of Transportation, National Highway Traffic Safety Administration, DOT HS 809 505, September 2002.

[6] Najm, W.G., Sen, B., Smith, J.D., and Campbell, B.N., *Analysis of Light Vehicle Crashes and Pre-Crash Scenarios Based on the 2000 General Estimates System*. U. S. Department of Transportation, National Highway Traffic Safety Administration, DOT HS 809 573, February 2003.

[7] Chovan, J.D., Tijerina, L., Alexandar, G., and Hendricks, D.L., *Examination of Lane Change Crashes and Potential IVHS Countermeasures*. U. S. Department of Transportation, National Highway Traffic Safety Administration, DOT HS 808 071, March 1994.

APPENDIX A. Identification of Dynamically Distinct Pre-Crash Scenarios Using GES Codes

This appendix describes the GES codes that were used to identify dynamically distinct pre-crash scenarios in rear-end, lane change, and run-off-road crashes for light vehicles and heavy trucks.

Vehicle Type Identification

List of Variables

1. BDYTYP_H = Hotdeck Imputed Body Type
2. SPEC_USE = Special Use

Codes

Light Vehicle
- BDYTYP_H = 01 – 22, 28 – 41, and 45 – 49

AND
- SPEC_USE = 00

Heavy Truck
- BDYTYP_H = 60, 64, 66, 78, and 79

Rear-End Pre-Crash Scenarios

List of Variables

1. ACC_TYPE = Accident Type
2. VROLE_I = Vehicle Role
3. TRAF_CON = Traffic Control Device
4. MANEUV_I = Univariate Imputed Movement Prior to Critical Event
5. P_CRASH2 = Critical Event

Prefix "v_" identifies the striking truck or light vehicle and prefix "ov_" refers to the other struck vehicle.

Codes

Crash population: ACC_TYPE = 20 – 43. Only two-vehicle crashes are considered.

Scenario 1: Striking (following) vehicle is making a maneuver (i.e., passing, leaving a parked position, entering a parked position, turning right, turning left, making U-turn, backing up, changing lanes, merging, corrective action, or other)
- v_VROLE_I = 1 AND v_MANEUV_I = 6, 8 – 13, 15 – 97

Scenario 2: Struck (lead) vehicle is accelerating

- ov_VROLE_I = 2 AND ov_MANEUV_I = 3 or 4

Scenario 3: Struck (lead) vehicle is moving at constant speed
- ov_ACC_TYPE = 25 – 27

OR
- ov_VROLE_I = 2 AND ov_MANEUV_I = 1 or 14

OR
- v_VROLE_I = 1 AND v_P_CRASH2 = 51

Scenario 4: Struck (lead) vehicle is decelerating
- ov_ACC_TYPE = 29 – 31

OR
- ov_VROLE_I = 2 AND ov_MANEUV_I = 2

OR
- ov_VROLE_I = 2 AND ov_P_CRASH2 = 18

OR
- v_VROLE_I = 1 AND v_P_CRASH2 = 52

Scenario 5: Struck (lead) vehicle is stopped in the process of turning (i.e., turning right, turning left, or making U-turn), or stopped in the presence of a traffic control device (i.e., trafficway traffic signal, stop sign, yield sign, officer, crossing guard, flagman, active devices at railroad grade crossing, or passive devices at railroad grade crossing)
- ov_ACC_TYPE = 21 – 23 AND ov_TRAF_CON = 1, 4, 8, 9, 21, 22, 51, 61, or 62

OR
- ov_ACC_TYPE = 21 – 23 AND ov_VROLE_I = 2 AND ov_MANEUV_I = 10 – 12

OR
- ov_ACC_TYPE = 21 – 23 AND ov_VROLE_I = 2 AND ov_P_CRASH2 = 15 or 16

OR
- ov_VROLE_I = 2 AND ov_MANEUV_I = 5 or 7 AND ov_TRAF_CON = 1, 4, 8, 9, 21, 22, 51, 61, or 62

OR
- ov_VROLE_I = 2 AND ov_MANEUV_I = 5 or 7 AND ov_P_CRASH2 = 15 or 16

OR
- v_VROLE_I = 1 AND v_P_CRASH2 = 50 AND v_TRAF_CON = 1, 4, 8, 9, 21, 22, 51, 61, or 62

OR
- v_VROLE_I = 1 AND v_P_CRASH2 = 50 AND ov_VROLE_I = 2 AND ov_MANEUV_I = 10 – 12

OR
- v_VROLE_I = 1 AND v_P_CRASH2 = 50 AND ov_VROLE_I = 2 AND ov_P_CRASH2 = 15 or 16

Scenario 6: Struck (lead) vehicle is stopped not in the process of turning and not in the presence of traffic control device
- ov_ACC_TYPE = 21 – 23

OR
- ov_VROLE_I = 2 AND ov_MANEUV_I = 5 or 7

OR
- v_VROLE_I = 1 AND v_P_CRASH2 = 50

Scenario 7: Other striking truck or light vehicle rear-end pre-crash scenarios
- v_ACC_TYPE = 20 – 43 AND v_VROLE_I = 1

Scenario 8: Other struck truck or light vehicle rear-end pre-crash scenarios

Lane Change Pre-Crash Scenarios

List of Variables

1. ACC_TYPE = Accident Type
2. MANEUV_I = Univariate Imputed Movement Prior to Critical Event
3. P_CRASH2 = Critical Event

Prefix "v_" identifies the truck or light vehicle changing lanes and prefix "ov_" refers to the other struck vehicle.

Codes

Crash population: ACC_TYPE = 44-49 and 70-73. Only two-vehicle crashes are considered.

Scenario 1: Vehicle changes lanes or passes to the right and encroaches on adjacent vehicle
- v_ACC_TYPE = 46 AND v_MANEUV_I = 6 or 15

OR
- v_MANEUV_I = 6 or 15 AND (v_P_CRASH2 = 11 OR ov_P_CRASH2 = 60)

Scenario 2: Vehicle changes lanes or passes to the left and encroaches on adjacent vehicle
- v_ACC_TYPE = 47 AND v_MANEUV_I = 6 or 15

OR
- v_MANEUV_I = 6 or 15 AND (v_P_CRASH2 = 10 OR ov_P_CRASH2 = 61)

Scenario 3: Vehicle is changing lanes or passing to unknown adjacent lane
- v_MANEUV_I = 6 or 15

Scenario 4: Vehicle merges to the right and encroaches on adjacent vehicle
- v_ACC_TYPE = 46 AND v_MANEUV_I = 16

OR
- v_MANEUV_I = 16 AND (v_P_CRASH2 = 11 OR ov_P_CRASH2 = 60)

Scenario 5: Vehicle merges to the left and encroaches on adjacent vehicle
- v_ACC_TYPE = 47 AND v_MANEUV_I = 16

OR
- v_MANEUV_I = 16 AND (v_P_CRASH2 = 10 OR ov_P_CRASH2 = 61)

Scenario 6: Vehicle is merging to unknown direction
- v_MANEUV_I = 16

Scenario 7: Vehicle turns to the right and encroaches on adjacent vehicle
- v_MANEUV_I = 10 OR v_P_CRASH2 = 16

Scenario 8: Vehicle turns to the left and encroaches on adjacent vehicle
- v_MANEUV_I = 11 OR v_P_CRASH2 = 15

Scenario 9: Vehicle drifts to the right and encroaches on adjacent vehicle
- v_MANEUV_I = 1, 2, 3, 4, or 14 AND (v_P_CRASH2 = 11 OR ov_P_CRASH2 = 60)

Scenario 10: Vehicle drifts to the left and encroaches on adjacent vehicle
- v_MANEUV_I = 1, 2, 3, 4, or 14 AND (v_P_CRASH2 = 10 OR ov_P_CRASH2 = 61)

Scenario 11: Vehicle is encroaching to adjacent lane on the right
- v_P_CRASH2 = 11 OR ov_P_CRASH2 = 60

Scenario 12: Vehicle is encroaching to adjacent lane on the left
- v_P_CRASH2 = 10 OR ov_P_CRASH2 = 61

Scenario 13: Other cases

Run-Off-Road Pre-Crash Scenarios

List of Variables

1. ACC_TYPE = Accident Type
2. MANEUV_I = Univariate Imputed Movement Prior to Critical Event
3. P_CRASH2 = Critical Event
4. ALIGN_I = Univariate Imputed Roadway Alignment

Codes

Crash population: ACC_TYPE = 01 – 16, excluding codes 03, 08, 13, and 14.

Scenario 1: Vehicle is going straight and departs road edge to the right
- MANEUV_I = 1 AND ALIGN_I = 1 AND P_CRASH2 = 11 or 13

Scenario 2: Vehicle is going straight and departs road edge to the left
- MANEUV_I = 1 AND ALIGN_I = 1 AND P_CRASH2 = 10 or 12

Scenario 3: Vehicle is negotiating a curve and departs road edge to the right

- MANEUV_I = 1 AND ALIGN_I = 2 AND P_CRASH2 = 11 or 13

OR
- MANEUV_I = 14 AND P_CRASH2 = 11 or 13

Scenario 4: Vehicle is negotiating a curve and departs road edge to the left
- MANEUV_I = 1 AND ALIGN_I = 2 AND P_CRASH2 = 10 or 12

OR
- MANEUV_I = 14 AND P_CRASH2 = 10 or 12

Scenario 5: Vehicle is going straight and loses control
- MANEUV_I = 1 AND ALIGN_I = 1 AND P_CRASH2 = 5 – 9

Scenario 6: Vehicle is negotiating a curve and loses control
- MANEUV_I = 1 AND ALIGN_I = 2 AND P_CRASH2 = 5 – 9

OR
- MANEUV_I = 14 AND P_CRASH2 = 5 – 9

Scenario 7: Vehicle is initiating a maneuver and loses control
- MANEUV_I = 2, 3, 4, 6, 8, 9, 10, 11, 12, 15, 16, 17, or 97 AND P_CRASH2 = 5 – 9

OR
- MANEUV_I = 2, 3, 4, 6, 8, 9, 10, 11, 12, 15, 16, 17, or 97 AND ACC_TYPE = 2 or 7 AND P_CRASH2 = 15, 16, or 18

Scenario 8: Vehicle is initiating a maneuver and departs road edge to the right
- MANEUV_I = 2, 3, 4, 6, 8, 9, 10, 11, 12, 15, 16, 17, or 97 AND P_CRASH2 = 11 or 13

OR
- MANEUV_I = 2, 3, 4, 6, 8, 9, 10, 11, 12, 15, 16, 17, or 97 AND ACC_TYPE = 1 AND P_CRASH2 = 15, 16, or 18

Scenario 9: Vehicle is initiating a maneuver and departs road edge to the left
- MANEUV_I = 2, 3, 4, 6, 8, 9, 10, 11, 12, 15, 16, 17, or 97 AND P_CRASH2 = 10 or 12

OR
- MANEUV_I = 2, 3, 4, 6, 8, 9, 10, 11, 12, 15, 16, 17, or 97 AND ACC_TYPE = 6 AND P_CRASH2 = 15, 16, or 18

Scenario 10: Other

APPENDIX B. Statistical Description of Base Test Scenarios

Light Vehicle Statistics Based on 2003 GES

Characteristics of Rear-End Crash Base Test Scenarios

Table 23. Driving Environment of Light Vehicle Rear-End Crash Base Test Scenarios

Base Scenario	Atmospheric Conditions				Road Alignment		Road Profile		
	Day Clear	Day Adverse	Dark Clear	Dark Adverse	Straight	Curve	Level	Slope	Hillcrest
A1	73%	8%	16%	3%	82%	18%	76%	23%	1%
A2	64%	10%	20%	6%	89%	11%	77%	20%	2%
A3	68%	11%	16%	4%	90%	10%	77%	21%	2%
A4	71%	12%	13%	4%	93%	7%	78%	20%	2%

Table 24. Speed Data of Light-Vehicle Rear-End Crash Base Test Scenarios

	A1			A2			A3			A4		
Speed (mph)	Speed Limit		Speeding	Speed Limit		Speeding	Speed Limit		Speeding	Speed Limit		Speeding
	Rel Freq	Cum Freq	Yes	Rel Freq	Cum Freq	Yes	Rel Freq	Cum Freq	Yes	Rel Freq	Cum Freq	Yes
<= 25	8%	8%	26%	9%	9%	29%	9%	9%	39%	8%	8%	37%
30	8%	17%	26%	8%	16%	20%	8%	17%	48%	9%	17%	46%
35	26%	42%	27%	20%	36%	36%	24%	41%	39%	26%	43%	29%
40	11%	53%	28%	10%	46%	28%	12%	54%	48%	13%	56%	38%
45	22%	74%	22%	18%	64%	25%	23%	76%	41%	21%	77%	33%
50	5%	80%	35%	5%	69%	19%	6%	82%	37%	4%	81%	27%
55	8%	88%	32%	13%	82%	33%	11%	93%	36%	10%	91%	28%
60	3%	91%	33%	6%	88%	44%	2%	95%	51%	2%	93%	60%
65	7%	97%	47%	9%	97%	62%	3%	98%	55%	5%	99%	52%
>= 70	3%	100%	25%	3%	100%	43%	2%	100%	61%	1%	100%	27%

Characteristics of Lane Change Crash Base Test Scenarios

Table 25. Driving Environment of Light-Vehicle Lane Change Crash Base Test Scenarios

Base Scenario	Atmospheric Conditions				Road Alignment		Road Profile			
	Day Clear	Day Adverse	Dark Clear	Dark Adverse	Straight	Curve	Level	Slope	Hillcrest	Sag
B1	67%	7%	23%	4%	92%	8%	81%	19%	1%	0%
B2	72%	8%	17%	4%	91%	9%	80%	19%	1%	0%
B3	68%	6%	18%	8%	98%	2%	84%	14%	1%	1%
B4	55%	12%	24%	9%	84%	16%	79%	19%	2%	0%

Table 26. Speed Data of Light-Vehicle Lane Change Crash Base Test Scenarios

Speed (mph)	B1			B2			B3			B4		
	Speed Limit		Speeding	Speed Limit		Speeding	Speed Limit		Speeding	Speed Limit		Speeding
	Rel Freq	Cum Freq	Yes	Rel Freq	Cum Freq	Yes	Rel Freq	Cum Freq	Yes	Rel Freq	Cum Freq	Yes
<= 25	7%	7%	4%	8%	8%	3%	21%	21%	1%	8%	8%	7%
30	9%	16%	0 2%	7%	15%	4%	14%	35%	0%	8%	16%	22%
35	24%	40%	2%	21%	36%	7%	28%	63%	0%	15%	31%	7%
40	12%	52%	5%	12%	48%	2%	7%	70%	0%	8%	40%	5%
45	17%	68%	6%	17%	65%	3%	21%	90%	0%	12%	52%	22%
50	8%	76%	7%	5%	70%	4%	3%	93%	0%	2%	54%	18%
55	9%	85%	14%	12%	81%	6%	6%	99%	0%	14%	68%	25%
60	4%	89%	24%	4%	86%	13%	1%	100%	0%	9%	77%	30%
65	8%	97%	7%	10%	96%	14%	0%	100%	0%	16%	93%	31%
>= 70	3%	100%	8%	4%	100%	6%	0%	100%	0%	7%	100%	44%

Characteristics of Run-Off-Road Crash Base Test Scenarios

Table 27. Driving Environment of Light-Vehicle Run-Off-Road Crash Base Test Scenarios

Base Scenario	Atmospheric Conditions				Road Alignment		Road Profile			
	Day Clear	Day Adverse	Dark Clear	Dark Adverse	Straight	Curve	Level	Slope	Hillcrest	Sag
C1	43%	7%	42%	7%	100%	0%	77%	22%	1%	0%
C2	35%	3%	54%	8%	100%	0%	76%	22%	1%	0%
C3	41%	9%	42%	9%	2%	98%	56%	41%	2%	1%
C4	31%	23%	31%	16%	2%	98%	51%	46%	3%	1%
C5	56%	5%	35%	4%	91%	9%	76%	22%	1%	0%

Table 28. Speed Data of Light-Vehicle Run-Off-Road Crash Base Test Scenarios

Speed (mph)	C1			C2			C3			C4			C5		
	Speed Limit		SpeedX	Speed Limit		SpeedX	Speed Limit		SpeedX	Speed Limit		SpeedX	Speed Limit		SpeedX
	Rel F	Cum F	Yes	Rel F	Cum F	Yes	Rel F	Cum F	Yes	Rel F	Cum F	Yes	Rel F	Cum F	Yes
<= 25	27%	27%	18%	18%	18%	11%	18%	18%	24%	16%	16%	74%	39%	39%	20%
30	11%	39%	16%	10%	28%	13%	7%	26%	23%	9%	25%	77%	16%	54%	31%
35	15%	53%	18%	11%	39%	15%	17%	43%	16%	17%	42%	82%	16%	70%	34%
40	6%	59%	12%	6%	45%	11%	6%	49%	27%	5%	47%	79%	5%	75%	25%
45	10%	68%	17%	9%	54%	10%	14%	63%	24%	14%	61%	84%	8%	83%	20%
50	2%	70%	16%	4%	57%	5%	2%	65%	30%	2%	63%	67%	2%	85%	0%
55	20%	91%	10%	22%	79%	10%	28%	93%	16%	27%	90%	68%	7%	92%	0%
60	1%	92%	35%	3%	82%	27%	2%	94%	4%	2%	92%	78%	2%	95%	31%
65	6%	98%	18%	12%	94%	23%	5%	99%	33%	5%	97%	67%	5%	99%	30%
>= 70	2%	100%	24%	6%	100%	26%	1%	100%	15%	3%	100%	73%	1%	100%	7%

Heavy-Truck Statistics Based on 2000-2003 GES

Characteristics of Rear-End Crash Base Test Scenarios

Table 29. Driving Environment of Heavy-Truck Rear-End Crash Base Test Scenarios

Base Scenario	Atmospheric Conditions				Road Alignment		Road Profile		
	Day Clear	Day Adverse	Dark Clear	Dark Adverse	Straight	Curve	Level	Slope	Hillcrest
D1	76%	8%	13%	3%	89%	11%	73%	27%	0%
D2	63%	8%	24%	5%	90%	10%	73%	24%	3%
D3	80%	7%	9%	4%	91%	9%	75%	23%	2%
D4	83%	8%	8%	1%	91%	9%	77%	22%	1%

Table 30. Speed Data of Heavy-Truck Rear-End Crash Base Test Scenarios

Speed (mph)	D1			D2			D3			D4		
	Speed Limit		Speeding	Speed Limit		Speeding	Speed Limit		Speeding	Speed Limit		Speeding
	Rel Freq	Cum Freq	Yes	Rel Freq	Cum Freq	Yes	Rel Freq	Cum Freq	Yes	Rel Freq	Cum Freq	Yes
<= 25	7%	7%	0%	3%	3%	28%	4%	4%	58%	7%	7%	0%
30	9%	17%	1%	3%	6%	33%	6%	10%	22%	8%	15%	0%
35	24%	41%	11%	10%	16%	2%	17%	27%	26%	22%	37%	7%
40	3%	44%	3%	3%	19%	4%	8%	35%	21%	10%	47%	0%
45	25%	69%	1%	14%	33%	23%	16%	52%	13%	19%	66%	0%
50	1%	70%	21%	3%	37%	11%	4%	55%	26%	5%	71%	0%
55	13%	83%	6%	24%	61%	11%	25%	80%	19%	20%	91%	10%
60	4%	88%	11%	6%	67%	27%	4%	84%	37%	3%	93%	11%
65	5%	93%	10%	12%	79%	44%	11%	95%	33%	4%	97%	58%
>= 70	7%	100%	58%	21%	100%	41%	5%	100%	29%	3%	100%	14%

Characteristics of Lane Change Crash Base Test Scenarios

Table 31. Driving Environment of Heavy-Truck Lane Change Crash Base Test Scenarios

Base Scenario	Atmospheric Conditions				Road Alignment		Road Profile		
	Day Clear	Day Adverse	Dark Clear	Dark Adverse	Straight	Curve	Level	Slope	Hillcrest
E1	72%	2%	19%	8%	94%	6%	80%	19%	1%
E2	76%	8%	13%	4%	88%	12%	73%	26%	0%
E3	82%	4%	13%	2%	96%	4%	86%	11%	2%
E4	77%	9%	14%	1%	91%	9%	76%	23%	2%

Table 32. Speed Data of Heavy-Truck Lane Change Crash Base Test Scenarios

Speed (mph)	E1 Speed Limit Rel Freq	E1 Speed Limit Cum Freq	E1 Speeding Yes	E2 Speed Limit Rel Freq	E2 Speed Limit Cum Freq	E2 Speeding Yes	E3 Speed Limit Rel Freq	E3 Speed Limit Cum Freq	E3 Speeding Yes	E4 Speed Limit Rel Freq	E4 Speed Limit Cum Freq	E4 Speeding Yes
<= 25	3%	3%	0%	4%	4%	0%	20%	20%	0%	8%	8%	0%
30	6%	8%	0%	2%	6%	0%	12%	32%	0%	5%	13%	0%
35	13%	21%	4%	10%	16%	0%	27%	59%	0%	25%	38%	0%
40	4%	25%	0%	10%	25%	0%	12%	71%	0%	3%	42%	2%
45	11%	36%	2%	8%	33%	0%	14%	85%	0%	13%	54%	0%
50	5%	42%	0%	0%	33%	6%	4%	88%	0%	6%	60%	0%
55	25%	67%	6%	28%	61%	1%	9%	97%	0%	18%	78%	11%
60	6%	73%	2%	6%	67%	1%	0%	97%	0%	6%	84%	3%
65	18%	91%	3%	18%	84%	0%	1%	98%	0%	11%	95%	20%
>= 70	9%	100%	7%	16%	100%	2%	2%	100%	0%	5%	100%	8%

Characteristics of Run-Off-Road Crash Base Test Scenarios

Table 33. Driving Environment of Heavy-Truck Run-Off-Road Crash Base Test Scenarios

Base Scenario	Atmospheric Conditions Day Clear	Day Adverse	Dark Clear	Dark Adverse	Road Alignment Straight	Curve	Road Profile Level	Slope	Hillcrest
F1	65%	5%	21%	8%	100%	0%	74%	26%	1%
F2	34%	6%	47%	14%	100%	0%	87%	13%	0%
F3	65%	8%	25%	2%	1%	99%	47%	53%	0%
F4	57%	16%	24%	3%	0%	100%	39%	60%	1%
F5	75%	8%	12%	5%	92%	8%	80%	18%	1%

Table 34. Speed Data of Heavy-Truck Run-Off-Road Crash Base Test Scenarios

Speed (mph)	F1 Speed Limit Rel F	F1 Speed Limit Cum F	F1 SpeedX Yes	F2 Speed Limit Rel F	F2 Speed Limit Cum F	F2 SpeedX Yes	F3 Speed Limit Rel F	F3 Speed Limit Cum F	F3 SpeedX Yes	F4 Speed Limit Rel F	F4 Speed Limit Cum F	F4 SpeedX Yes	F5 Speed Limit Rel F	F5 Speed Limit Cum F	F5 SpeedX Yes
<= 25	26%	26%	5%	26%	26%	0%	19%	19%	8%	7%	7%	0%	48%	48%	2%
30	6%	33%	2%	10%	36%	35%	3%	23%	0%	2%	9%	0%	10%	58%	10%
35	11%	44%	0%	2%	38%	4%	7%	29%	24%	15%	24%	78%	19%	78%	1%
40	1%	45%	7%	0%	38%	0%	5%	34%	0%	5%	29%	0%	7%	84%	0%
45	8%	53%	30%	12%	50%	9%	14%	48%	40%	16%	45%	0%	5%	90%	1%
50	2%	55%	56%	0%	50%	34%	1%	49%	83%	4%	49%	0%	1%	91%	0%
55	24%	79%	6%	19%	69%	13%	46%	95%	24%	32%	82%	58%	7%	98%	2%
60	1%	80%	2%	1%	70%	12%	1%	96%	53%	3%	85%	61%	0%	98%	25%
65	9%	89%	17%	7%	77%	21%	2%	98%	52%	7%	92%	78%	1%	99%	1%
>= 70	11%	100%	17%	23%	100%	62%	2%	100%	44%	8%	100%	27%	1%	100%	15%

APPENDIX C. First Harmful Event Statistics of Run-Off-Road Crash Scenarios

Light-Vehicle Statistics Based on 2003 GES

Table 35. Light-Vehicle First Harmful Events: Going Straight and Departing Road Edge Scenarios

First Harmful Event	Right Road Departure		Left Road Departure		Both Scenarios	
	Frequency	Rel. Freq.	Frequency	Rel. Freq.	Frequency	Rel. Freq.
Parked vehicle	64,000	36.0%	15,000	18.7%	80,000	30.6%
Post, pole, or support	33,000	18.4%	14,000	17.6%	47,000	18.2%
Culvert or ditch	18,000	10.2%	9,000	10.5%	27,000	10.3%
Tree	12,000	6.5%	8,000	10.3%	20,000	7.7%
Guardrail	9,000	4.8%	7,000	8.3%	15,000	5.9%
Rollover/overturn	9,000	4.9%	5,000	6.5%	14,000	5.4%
Other fixed object	6,000	3.5%	3,000	3.5%	9,000	3.5%
Embankment	6,000	3.4%	3,000	3.5%	9,000	3.4%
Fence	5,000	2.8%	4,000	4.5%	9,000	3.3%
Traffic barrier	2,000	0.9%	6,000	7.7%	8,000	3.1%
Curb	6,000	3.4%	2,000	2.4%	8,000	3.0%
Bridge structure	2,000	1.1%	1,000	1.7%	3,000	1.3%
Other object not fixed	2,000	1.0%	1,000	1.1%	3,000	1.1%
Fire hydrant	2,000	0.9%	-	0.6%	2,000	0.8%
Shrubbery or bush	1,000	0.7%	1,000	0.8%	2,000	0.7%
Wall	1,000	0.6%	-	0.6%	2,000	0.6%
Building	-	0.3%	1,000	0.8%	1,000	0.4%
Boulder	-	0.2%	-	0.4%	1,000	0.3%
Ground	-	0.2%	-	0.1%	1,000	0.2%
Impact attenuator	-	0.1%	-	0.2%	-	0.1%
Fixed object - no details	-	0.1%	-	0.1%	-	0.1%
Pedestrian	-	0.04%	-	0.03%	-	0.04%
Total	179,000	100.0%	82,000	100.0%	261,000	100.0%

Table 36. Light-Vehicle First Harmful Events: Negotiating a Curve and Departing Road Edge Scenarios

First Harmful Event	Right Road Departure		Left Road Departure		Both Scenarios	
	Frequency	Rel. Freq.	Frequency	Rel. Freq.	Frequency	Rel. Freq.
Post, pole, or support	15,000	20.8%	6,000	14.5%	22,000	18.5%
Tree	10,000	13.9%	6,000	15.2%	17,000	14.4%
Culvert or ditch	11,000	14.7%	5,000	11.1%	16,000	13.4%
Parked vehicle	6,000	8.6%	4,000	8.8%	10,000	8.6%
Guardrail	6,000	7.6%	4,000	9.8%	10,000	8.4%
Rollover/overturn	6,000	7.8%	3,000	7.6%	9,000	7.7%
Embankment	5,000	7.0%	3,000	7.8%	9,000	7.3%
Curb	3,000	3.5%	3,000	7.7%	6,000	5.0%
Traffic barrier	1,000	1.7%	3,000	7.6%	5,000	3.9%
Fence	3,000	3.8%	1,000	2.3%	4,000	3.3%
Other fixed object	3,000	3.8%	1,000	1.8%	4,000	3.1%
Boulder	1,000	1.1%	1,000	2.2%	2,000	1.5%
Other object not fixed	1,000	1.0%	-	1.0%	1,000	1.0%
Shrubbery or bush	1,000	1.0%	-	0.5%	1,000	0.8%
Wall	1,000	1.0%	-	0.5%	1,000	0.8%
Bridge structure	1,000	1.1%	-	0.1%	1,000	0.7%
Building	-	0.4%	-	0.5%	1,000	0.4%
Ground	-	0.5%	-	0.03%	-	0.3%
Fixed object - no details	-	0.3%	-	0.2%	-	0.3%
Fire hydrant	-	0.3%	-	0.2%	-	0.3%
Impact attenuator	-	0.01%	-	0.4%	-	0.2%
Cyclist	-	0.03%	-	0.0%	-	0.02%
Total	74,000	100.0%	42,000	100.0%	116,000	100.0%

Table 37. Light-Vehicle First Harmful Events: Control Loss Scenarios

First Harmful Event	Going Straight		Negotiating a Curve		Both Scenarios	
	Frequency	Rel. Freq.	Frequency	Rel. Freq.	Frequency	Rel. Freq.
Post, pole, or support	36,000	17.5%	25,000	14.3%	61,000	16.0%
Tree	28,000	13.7%	30,000	17.3%	58,000	15.3%
Culvert or ditch	26,000	12.5%	21,000	12.2%	47,000	12.3%
Guardrail	25,000	11.9%	22,000	12.8%	47,000	12.3%
Rollover/overturn	21,000	10.3%	13,000	7.4%	34,000	9.0%
Traffic barrier	20,000	9.4%	9,000	5.1%	28,000	7.5%
Embankment	7,000	3.3%	15,000	8.7%	22,000	5.7%
Parked vehicle	9,000	4.4%	5,000	3.0%	14,000	3.7%
Fence	6,000	3.0%	6,000	3.8%	13,000	3.3%
Curb	7,000	3.1%	5,000	2.9%	12,000	3.0%
Other fixed object	6,000	2.7%	6,000	3.2%	11,000	2.9%
Bridge structure	7,000	3.3%	3,000	1.5%	10,000	2.5%
Other object not fixed	4,000	1.7%	4,000	2.3%	8,000	2.0%
Wall	2,000	0.8%	3,000	1.9%	5,000	1.3%
Building	1,000	0.6%	1,000	0.5%	2,000	0.6%
Fire hydrant	1,000	0.4%	1,000	0.7%	2,000	0.5%
Shrubbery or bush	1,000	0.4%	1,000	0.7%	2,000	0.5%
boulder	-	0.2%	1,000	0.8%	2,000	0.5%
Fixed object - no details	1,000	0.4%	1,000	0.4%	2,000	0.4%
Ground	-	0.2%	-	0.2%	1,000	0.2%
Other noncollision	-	0.0%	-	0.1%	-	0.1%
Impact attenuator	-	0.0%	-	0.1%	-	0.1%
Potholes	-	0.1%	-	0.0%	-	0.1%
Pedestrian	-	0.01%	-	0.1%	-	0.0%
Animal	-	0.1%	-	0.0%	-	0.0%
Object not fixed - no details	-	0.04%	-	0.0%	-	0.0%
Railway train	-	0.04%	-	0.0%	-	0.0%
Total	208,000	100.0%	172,000	100.0%	380,000	100.0%

Table 38. Light-Vehicle First Harmful Events: Initiating a Maneuver Scenarios

First Harmful Event	Control Loss		Right Road Departure		Left Road Departure		Three Scenarios	
	Frequency	Rel. Freq.	Frequency	Rel. Freq.	Frequency	Rel. Freq.	Frequency	Rel. Freq.
Parked vehicle	5,000	8.5%	18,000	42.9%	5,000	23.9%	28,000	23.5%
Post, pole, or support	15,000	27.7%	7,000	15.5%	5,000	20.2%	27,000	22.0%
Curb	4,000	8.0%	4,000	8.6%	3,000	12.6%	11,000	9.1%
Tree	6,000	10.1%	2,000	4.1%	1,000	5.0%	8,000	7.0%
Guardrail	5,000	9.6%	2,000	4.7%	1,000	3.1%	8,000	6.6%
Culvert or ditch	4,000	7.1%	2,000	5.5%	2,000	7.6%	8,000	6.6%
Traffic barrier	4,000	7.9%	1,000	2.6%	2,000	7.4%	7,000	6.0%
Fence	2,000	4.3%	2,000	4.0%	1,000	3.4%	5,000	4.0%
Other fixed object	2,000	3.4%	1,000	2.8%	-	2.1%	4,000	3.0%
Rollover/overturn	2,000	3.9%	-	1.0%	1,000	2.7%	3,000	2.6%
Embankment	1,000	2.3%	1,000	1.9%	-	0.5%	2,000	1.8%
Building	-	0.9%	-	0.6%	1,000	3.4%	2,000	1.3%
Fire hydrant	1,000	1.7%	-	0.5%	-	0.9%	1,000	1.2%
Other object not fixed	1,000	1.4%	-	1.2%	-	0.5%	1,000	1.1%
Wall	1,000	1.1%	-	0.8%	-	1.0%	1,000	1.0%
Impact attenuator	-	0.2%	1,000	1.4%	-	1.9%	1,000	0.9%
Bridge structure	-	0.5%	-	0.4%	-	1.1%	1,000	0.6%
Fixed object - no details	-	0.7%	-	0.0%	-	0.5%	1,000	0.4%
Shrubbery or bush	-	0.6%	-	0.3%	-	0.0%	-	0.4%
Boulder	-	0.2%	-	0.5%	-	0.5%	-	0.3%
Pedestrian	-	0.0%	-	0.2%	-	1.1%	-	0.3%
Ground	-	0.0%	-	0.6%	-	0.0%	-	0.2%
Animal	-	0.0%	-	0.0%	-	0.5%	-	0.1%
Other type non-motorist	-	0.0%	-	0.0%	-	0.2%	-	0.04%
Total	55,000	100.0%	42,000	100.0%	23,000	100.0%	121,000	100.0%

Heavy Truck Statistics Based on 2000-2003 GES

Table 39. Heavy-Truck First Harmful Events: Going Straight and Departing Road Edge Scenarios

First Harmful Event	Right Road Departure			Left Road Departure			Both Scenarios		
	Cases	Frequency	Rel. Freq.	Cases	Frequency	Rel. Freq.	Cases	Frequency	Rel. Freq.
Parked Motor Vehicle	137	14,590	51.8%	14	1,961	26.3%	151	16,551	46.5%
Rollover/Overturn	73	2,945	10.5%	28	1,350	18.1%	101	4,296	12.1%
Post, Pole or Support	42	2,547	9.0%	14	787	10.5%	56	3,335	9.4%
Guardrail	65	2,606	9.3%	23	726	9.7%	88	3,332	9.4%
Culvert or Ditch	55	1,818	6.5%	16	632	8.5%	71	2,450	6.9%
Other Fixed Object	8	672	2.4%	2	278	3.7%	10	950	2.7%
Fence	12	714	2.5%	4	153	2.0%	16	867	2.4%
Embankment	13	595	2.1%	7	177	2.4%	20	772	2.2%
Tree	22	638	2.3%	9	116	1.6%	31	754	2.1%
Other Object Not Fixed	2	332	1.2%	1	277	3.7%	3	609	1.7%
Shrubbery or Bush	1	7	0.0%	3	387	5.2%	4	394	1.1%
Bridge Structure	8	214	0.8%	1	55	0.7%	9	269	0.8%
Traffic Barrier	5	66	0.2%	19	164	2.2%	24	230	0.6%
Curb	7	79	0.3%	1	80	1.1%	8	158	0.4%
Building	4	68	0.2%	2	86	1.1%	6	154	0.4%
Fixed Object, No Details	1	57	0.2%	1	95	1.3%	2	152	0.4%
Impact Attenuator	6	89	0.3%	3	14	0.2%	9	103	0.3%
Ground	1	87	0.3%	-	-	0%	1	87	0.2%
Wall	2	14	0.05%	3	70	0.9%	5	84	0.2%
Boulder	1	13	0.05%	1	55	0.7%	2	69	0.2%
Fire Hydrant	1	5	0.02%	-	-	0%	1	5	0.01%
Total	466	28,156	100%	152	7,464	100%	618	35,620	100%

Table 40. Heavy-Truck First Harmful Events: Negotiating a Curve and Departing Road Edge Scenarios

First Harmful Event	Right Road Departure			Left Road Departure			Both Scenarios		
	Cases	Frequency	Rel. Freq.	Cases	Frequency	Rel. Freq.	Cases	Frequency	Rel. Freq.
Guardrail	34	2,235	20.5%	14	83	3.9%	48	2,318	17.8%
Rollover/Overturn	35	1,325	12.2%	33	478	22.7%	68	1,803	13.9%
Parked Motor Vehicle	15	1,386	12.7%	2	302	14.3%	17	1,688	13.0%
Post, Pole or Support	23	883	8.1%	8	557	26.4%	31	1,439	11.1%
Embankment	10	1,232	11.3%	5	88	4.2%	15	1,321	10.2%
Culvert or Ditch	13	852	7.8%	8	199	9.4%	21	1,051	8.1%
Fence	7	708	6.5%	2	44	2.1%	9	752	5.8%
Curb	7	393	3.6%	8	42	2.0%	15	435	3.3%
Traffic Barrier	8	328	3.0%	10	105	5.0%	18	433	3.3%
Other Fixed Object	8	354	3.2%	1	20	0.9%	9	373	2.9%
Wall	4	327	3.0%	3	41	2.0%	7	368	2.8%
Bridge Structure	5	285	2.6%	3	21	1.0%	8	306	2.4%
Tree	5	256	2.4%	-	-	0%	5	256	2.0%
Ground	1	246	2.3%	-	-	0%	1	246	1.9%
Other Object Not Fixed	2	42	0.4%	1	52	2.5%	3	94	0.7%
Impact Attenuator	2	10	0.1%	1	79	3.7%	3	89	0.7%
Boulder	1	22	0.2%	-	-	0%	1	22	0.2%
Fire Hydrant	1	5	0.04%	-	-	0%	1	5	0.04%
Total	181	10,890	100%	99	2,111	100%	280	13,001	100%

Table 41. Heavy-Truck First Harmful Events: Control Loss Scenarios

First Harmful Event	Going Straight			Negotiating a Curve			Both Scenarios		
	Cases	Frequency	Rel. Freq.	Cases	Frequency	Rel. Freq.	Cases	Frequency	Rel. Freq.
Rollover/Overturn	54	1,506	21.0%	152	3,207	34.5%	206	4,713	28.6%
Guardrail	58	1,245	17.4%	67	1,648	17.7%	125	2,893	17.6%
Culvert or Ditch	21	834	11.6%	22	544	5.9%	43	1,378	8.4%
Post, Pole or Support	21	744	10.4%	23	541	5.8%	44	1,285	7.8%
Parked Motor Vehicle	12	618	8.6%	6	561	6.0%	18	1,178	7.2%
Embankment	9	309	4.3%	13	827	8.9%	22	1,136	6.9%
Traffic Barrier	56	637	8.9%	35	174	1.9%	91	811	4.9%
Fence	6	163	2.3%	3	535	5.8%	9	698	4.2%
Tree	15	247	3.4%	19	405	4.4%	34	651	4.0%
Curb	5	298	4.2%	8	40	0.4%	13	339	2.1%
Bridge Structure	8	103	1.4%	9	218	2.4%	17	322	2.0%
Wall	6	33	0.5%	11	263	2.8%	17	296	1.8%
Building	-	-	0%	1	243	2.6%	1	243	1.5%
Fixed Object, No Details	1	162	2.3%	-	-	0%	1	162	1.0%
Other Fixed Object	4	66	0.9%	2	62	0.7%	6	127	0.8%
Other Object Not Fixed	2	85	1.2%	-	-	0%	2	85	0.5%
Impact Attenuator	4	48	0.7%	1	5	0.1%	5	52	0.3%
Ground	2	30	0.4%	1	17	0.2%	3	47	0.3%
Fire Hydrant	3	38	0.5%	-	-	0%	3	38	0.2%
Pavement Irregularity	1	2	0.03%	-	-	0%	1	2	0.01%
Total	288	7,167	100%	373	9,290	100%	661	16,457	100%

Table 42. Heavy-Truck First Harmful Events: Initiating a Maneuver Scenarios

First Harmful Event	Control Loss			Right Road Departure			Left Road Departure			All Three Scenarios		
	Cases	Frequency	Rel. Freq.	Cases	Frequency	Rel. Freq.	Cases	Frequency	Rel. Freq.	Cases	Frequency	Rel. Freq.
Parked Motor Vehicle	8	1,075	21.1%	64	13,162	36.8%	14	2,344	43.2%	86	16,581	35.8%
Post, Pole or Support	16	1,119	22.0%	62	12,924	36.2%	10	1,502	27.7%	88	15,545	33.6%
Rollover/Overturn	45	1,404	27.6%	26	766	2.1%	7	437	8.1%	78	2,607	5.6%
Guardrail	13	368	7.2%	15	978	2.7%	7	607	11.2%	35	1,953	4.2%
Fire Hydrant	-	-	0.0%	6	1,411	3.9%	1	267	4.9%	7	1,677	3.6%
Other Fixed Object	1	263	5.2%	7	1,377	3.9%	2	14	0.3%	10	1,653	3.6%
Culvert or Ditch	4	308	6.1%	10	1,046	2.9%	2	54	1.0%	16	1,407	3.0%
Building	-	-	0.0%	4	1,089	3.0%	-	-	0.0%	4	1,089	2.4%
Fence	1	259	5.1%	3	742	2.1%	1	5	0.1%	5	1,007	2.2%
Curb	5	43	0.8%	7	817	2.3%	1	5	0.1%	13	866	1.9%
Tree	4	43	0.8%	6	692	1.9%	1	3	0.1%	11	739	1.6%
Traffic Barrier	15	61	1.2%	3	250	0.7%	5	111	2.0%	23	423	0.9%
Impact Attenuator	2	9	0.2%	2	276	0.8%	1	5	0.1%	5	290	0.6%
Bridge Structure	2	27	0.5%	1	211	0.6%	4	15	0.3%	7	252	0.5%
Ground	-	-	0%	-	-	0%	1	55	1.0%	1	55	0.1%
Other Object Not Fixed	1	34	0.7%	1	6	0.02%	-	-	0%	2	40	0.1%
Boulder	2	25	0.5%	-	-	0%	-	-	0%	2	25	0.1%
Embankment	5	21	0.4%	-	-	0%	-	-	0%	5	21	0.05%
Wall	2	12	0.2%	-	-	0%	1	5	0.1%	3	18	0.04%
Shrubbery or Bush	2	17	0.3%	-	-	0%	-	-	0%	2	17	0.04%
Total	128	5,090	100%	217	35,748	100%	58	5,429	100%	403	46,267	100%

DOT HS 810 757
April 2007

www.ingramcontent.com/pod-product-compliance
Lightning Source LLC
Chambersburg PA
CBHW081900170526
45167CB00007B/3085